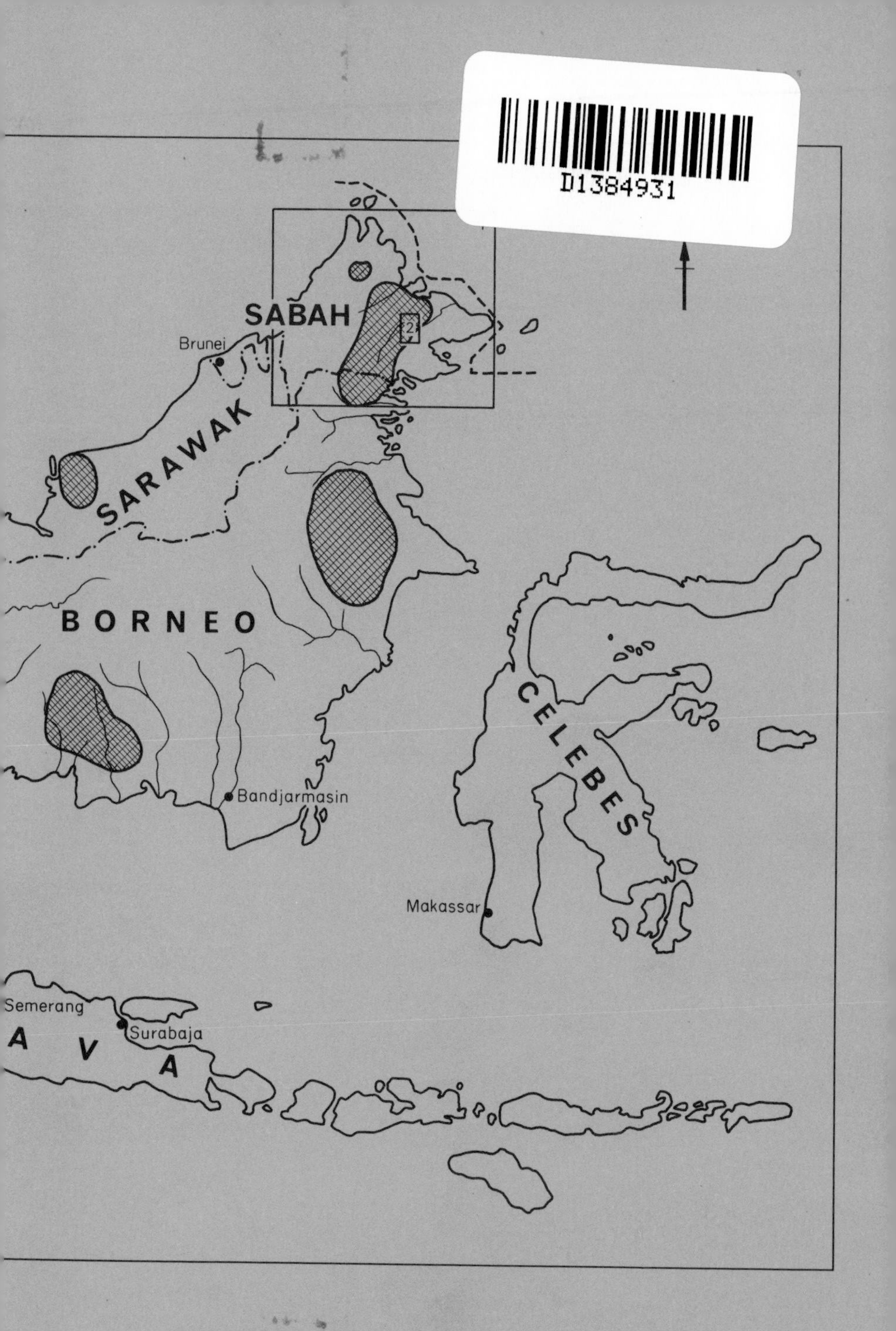

SABAH
Brunei
SARAWAK
BORNEO
Bandjarmasin
CELEBES
Makassar
Semerang
Surabaja
A V A

In Search of The Red Ape

In Search of The Red Ape

John MacKinnon

COLLINS
St James's Place, London, 1974

William Collins Sons & Co Ltd
London · Glasgow · Sydney · Auckland
Toronto · Johannesburg

First published 1974

ISBN 0 00 216703 4

Set in Baskerville
Made and Printed in Great Britain by
William Collins Sons & Co Ltd Glasgow

For Kathy and Jamie
my long-suffering family

Contents

Illustrations

All photographs were taken by the author – except for the Red-crested Partridge, taken by R. Parker (page 121) and the author and son, taken by H. Rijksen (page 201). The author is grateful to Mr Parker and Mr Rijksen for kind permission to use their photographs.

Maps

Foreword by

David Attenborough

The orang-utan was for long the most mysterious of the world's three great apes. Years after Jane Goodall had lived with a chimpanzee troop and George Schaller had watched gorillas for hundreds of hours, the orang-utan remained virtually unknown in the wild.

And for good reason. It lives not in troops or families but in groups of two or three, or solitarily. Nor does it have readily identifiable territorial habits. Even finding one at all is far from easy, as I discovered on a journey in Borneo many years ago; and once located, there is no way of ensuring that you can find it again, so unpredictable are its wanderings. In short, studying it for long periods of time in the wild might well be regarded as a near-impossibility by any scientist who requires a guaranteed minimum of creature comforts in order to keep body and soul together, let alone take notes at the same time.

John MacKinnon's solution to this problem had an appalling starkness. He decided that once he found an individual, he would simply follow it through the forest, sleeping where it slept, moving when it moved. What is more, he would do so for days and if necessary weeks on end; and he would take with him no more than the minimum rations, a knife and a sheet of plastic under which to sleep.

The dangers of the jungle to a conventional traveller are not those painted traditionally by Hollywood and fictional adventure. Venomous snakes do not dangle from every branch, nor do ravening carnivores lurk in ambush behind every tree. In my experience, by far the gravest hazard is that of getting hopelessly lost. Animals seldom even threaten. Indeed, you usually have to be lucky to catch a glimpse of one.

But paradoxically, the dangers are greatly increased if you become as skilled a jungle traveller as John MacKinnon. Having spent some time with him in the forests of Sumatra, I know how silently he can move through the thickest jungle. It is only such a man who can – either by design or, even more dangerously, by accident – get close to a forest elephant and therefore be likely to provoke its charge. And only someone as stoically hardy as he would inadvertently make his bed on the regular path of a porcupine, simply because an ape had chosen to go to sleep in the branches a hundred feet above that particular spot the previous evening.

The discoveries about the orang-utan that John MacKinnon made as a result of these extraordinary journeys are of great scientific importance and he writes of many of them here. But for me, this book is also particularly memorable because it is a vivid record of as close and intimate an association between a naturalist and the magical forests of the Far East, as I know.

DAVID ATTENBOROUGH

Part One

Borneo 1968

Chapter one

The Adventure Begins

The little bamboo hut crouches almost unseen in the dark forest. Over a glowing log the blackened pots steam ceaselessly. Wisps of smoke filter through the grass roof and the aromatic smell fills me with pleasing memories of other fires and other nights.

I can hear the roar of a great waterfall but it is far away and does not hide the other rich sounds of the jungle night; the hum of frogs and crickets, the argus pheasant plaintively calling to its mate and the harsh squeals of the macaque monkeys squabbling over sleeping arrangements in the old fig-tree down by the river. Farther away, curled up in a leafy treetop nest, is a bundle of shaggy red hair. Sometimes he, too, joins in the night chorus, giving a long series of deep and terrifying roars, but to-night he is content and quiet. The weather has been fine for a week and he has just enjoyed a great feast of wild mangosteens.

Scientists and settlers alike have argued over what to call this 'Man of the Woods' since his official 'discovery' three centuries ago. To-day he is usually referred to as 'orang-utan', but to the people living at the edge of the forest he has been known simply as '*Mawas*' since the days when he first broke their puny Stone Age spears and became revered in their legends as the strongest animal in the forest.

'No animal is strong enough to hurt the *Mawas*, and the only creature he ever fights with is the crocodile. When there is no fruit in the jungle, he goes to seek food on the banks of the river, where there are plenty of young shoots that he likes, and fruits that grow close to the water. Then the crocodile sometimes tries to seize him, but the *Mawas* gets upon him, and beats him with his hands and feet, and tears him and kills him . . . He always kills the crocodile by main strength, standing upon it, pulling open its jaws, and ripping up its throat . . . The *Mawas* is very strong; there is no animal in the jungle so strong as he.'[1]

Other tales tell of the orang-utan's origins. One theory claims that the orang-utan is descended from a man who, ashamed of a misdeed in the village, fled into the forest and remained there. Another story tells of the two bird-like creatures who were the creators of all life. They made many types of animal but when they made man and woman they were so pleased with themselves that they held a great feast. The next day they returned to make some more people but their excesses of the previous night had made them forget an important ingredient and they succeeded in making only orang-utans.

Like the African apes and South American monkeys the orang is often accused of sexual assault upon humans. A common fable tells of an orang-utan that snatched a young native girl and took her up into the trees where he kept her captive and fed her on fruit. Eventually she had a baby that was half man, half ape. One day when the orang-utan was not looking she escaped, climbing down a rope she had made of coconut fibre. She ran through the forest clutching her baby but the orang came swinging through the trees after her. She made for the river and saw a boat just about to leave the shore. The orang-utan nearly caught her but

[1] From A. R. Wallace, *The Malay Archipelago*, 1869

the men in the boat cried to her to drop the child. This she did and the ape stopped to seize the baby, giving her time to reach the boat and get away. In rage the orang-utan tore the baby in half and hurled the human half after the departing boat, the ape half he threw back into the forest. The loud calls of the orang-utan are said to be his sighs as he still searches for his lost bride. Other stories even tell of young men being abducted by female orang-utans. Nor are such fables confined to the native peoples of Borneo and Sumatra, where the apes occur. Europeans have added their own accounts of foul deeds, as in Edgar Allan Poe's *The Murders in the Rue Morgue*, or damsels abducted and rescued in the nick of time, as in O. C. Vane's story, *To the Rescue*.

The man-like appearance of the ape and the wealth of legend that surrounds him have long captured the imagination of western man yet the *Mawas* has remained shrouded in mystery as attempts to discover its habits in the wild have yielded little or no new information. This is now my quest and for three years I have followed the *Mawas* through the forests of Borneo and Sumatra. For him I have become a resident in the timeless world of the virgin jungle, where all the seasons are alike.

In a small clearing my three native helpers have built our small house, a raised bamboo platform, open at the front and with a grass roof and sides. The three men hired from the nearest village run the camp and tend to my needs while I tramp the forest in search of orang-utan. The distinguished Nami, an elder of the village, looks after the tapioca, maize and fiery chilli peppers that grow in the clearing and feeds the two pigs and chickens that we have brought. Genial Purba, cousin of the local policeman, collects firewood, cleans the camp and cooks for us. Once a fortnight he makes a seven-hour trek to the nearest market to buy fruit and collect mail and supplies. Turut, small and wizened, knows the habits of the jungle fowl and fish. He is not in my regular employ but gets paid for the meat he brings into our camp. Certainly the odd eel or pheasant is very welcome in our pot. Fifty yards down the hill is the narrow coffee-coloured river that serves as our bath as well as our drinking water.

All around us the mountains rise, cloaked in exotic vegetation. Giant buttressed trees and twisting lianas on the lower slopes give way to strange moss-bearded trees and insect eating pitcher-plants above. This is the home of the elephant and the tiger, the rhinoceros and the wild boar. Beside the river live the crocodile and the python and in the hills wild goats and hunting dogs. At night the panther and scorpion stalk their prey but during the day dazzling birds flit and twitter through the treetops. Yet despite this great wealth of plants and animals, my helpers and I are alone here for man is ill-adapted to live in such a place. While cousin *Mawas* rests content in his nest the men exchange terrifying tales of wandering spirits and ponder on the madness of the Englishman and his strange quest.

At times I, too, question my motives. I recall the biting wind and swirling fog of the winter's day on Dartmoor when I first conceived this study. Perhaps it was the very cold of that occasion that turned my mind to humid tropical forests, but my interest in apes had developed earlier than this. Before going up to university I had the luck to spend a fascinating and instructive year, working with Jane and Hugo van Lawick at the Gombe Stream Reserve in Tanzania. I originally intended to study only insects there but who could resist the excitable chimpanzees and their engaging personalities? It was in Gombe that I learned the techniques of tracking animals and observing and recording the behaviour of wild primates. When I returned to Oxford I took with me a deep interest in apes.

Excellent pioneer field-studies had already been carried out on the gorilla and gibbon, but the other ape, the orang-utan, remained something of an enigma. Several expeditions had attempted to learn about this rare and elusive creature but, apart from Barbara Harrisson's work with captive animals, little was known about its behaviour and social organisation in the wild. Equipped with the knowledge and experience from my chimp days and a good deal of youthful optimism, I felt sure that I could be more successful.

The main problems in studying orang-utans were the rarity and shyness of the apes and the inhospitality of the jungle. In contrast to previous expeditions I decided to travel alone with only a minimum of equipment, since the added mobility and quietness should enable me to encounter more animals and they would be less shy of a solitary observer. It would also be necessary to sleep out in the forest to maintain contact with the orangs through the night. These plans aroused alarm and incredulity. People pointed out the dangers from elephants, bears, snakes and falling trees, the risk of getting lost, sickness, disease or injury with no possible rescue and the discomfort of the heat, rain, mud, thorns, biting flies and blood-sucking leeches. My mind, however, was made up. I was full of confidence and very stubborn. There was much to be done seeking sponsorship and funds for the trip, deciding on a site and getting permission to conduct the study and gathering supplies and equipment.

Due to serious poaching and hunting pressure the range of the orang-utan had been drastically reduced and the ape was now confined to a few small areas in Borneo and North Sumatra. Since in 1967 Indonesia was still unsettled after her recent revolution and confrontation I decided to work in Malaysian Borneo. In Sarawak the few remaining orang strongholds were all in prohibited security zones near the Kalimantan border. In Sabah, however, several reports suggested that the orang was not uncommon in the forests around the eastern rivers. Eventually, after many letters to and fro, I received permission to work there.

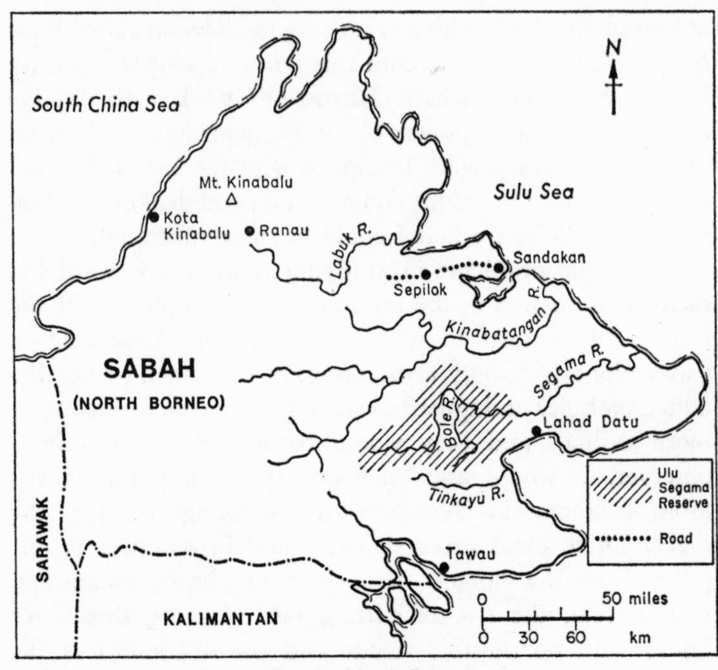

Sabah, North Borneo

Money was hard to find. Few people seemed willing to back such a crazy scheme, but something of the madness must have tickled the fancy of the Oxford University Expeditionary Council for I received full recognition and support from them. Several funds promised small sums but the trip would be expensive and it seemed unlikely that I would ever reach my minimal target. Just when it looked as though I would have to abandon my plans, I heard that further funds had come from France. Overjoyed at this news, I immediately booked my flights to the Far East and rushed about making last minute arrangements. After a final visit to my doctor I staggered from the surgery, pumped full of inoculations and clutching two large carrier bags filled with medicaments for every conceivable illness I might encounter on the trip. Half an hour later a sympathetic friend found me squatting on a pavement rummaging through my exotic drugs. Unfortunately there was nothing to relieve the pain in my rear which had seized up from one of the injections.

At last I was on my way, flying over the north coast of Borneo after a three-day stop in Singapore, where I had been able to buy a camera and absorb some of the oriental atmosphere. Between glances at the book of Malay grammar on my knee, I gazed down at the endless green jungle and twisting brown rivers that were to be my home for the next few months. To my left the rocky mass of Mount Kinabalu towered above the surrounding hills and in a valley below the rooftops of Ranau glinted in the sunshine. From this height the forest looked innocent enough but I remembered the thousands of Australian prisoners who had died there, on the terrible march from the coast, only twenty-three years earlier.

In Sandakan I discussed my plans with the game warden, Stanley de Silva. Together we pored over a large map of the Ulu Segama Reserve; 600 square miles of virgin forest where no zoologist had ever been. The region could only be reached by small boat up the treacherous Segama River, but I could hire men and boats at the coastal town of Lahad Datu.

Stanley then told me about his own work, a rehabilitation scheme for captive orang-utans in the nearby Sepilok Reserve. Illegally owned animals are confiscated by the Forest Department and released at the jungle camp. Here they are sheltered and fed but are completely free to wander where they will. Since most of the animals have been removed from their mothers at an early age, they have had no chance to learn the vital skills needed to survive in the forest. At Sepilok they are encouraged to follow the more experienced animals into the jungle. Those that are reluctant to leave camp are carried piggy-back by game-rangers far into the forest and left to find their own way home. Gradually by imitation and discovery the animals learn to be self-sufficient so that it is hoped they can eventually be released in other parts of the country to boost declining wild populations.

I decided to visit the camp myself and travelled the next day on a jaunty colourful bus, packed tight with grinning brown natives and squawking chickens. The bus bumped on along the rutted road through plantations of rubber and *kampongs* of stilted houses and leafy banana trees. At stop after stop the passengers climbed down till only I was left. The bus halted, the driver pointed to a narrow path and I set off on the short walk to the camp.

Six eager red apes came to greet me. I was gripped by strong hands and feet while they examined every part of my clothing and anatomy. These formalities completed I was able to take a better look at my new friends. The bald clown, Jippo, somersaulted in rings around me, hugging himself with jealousy when I paid too much attention to the more introverted and supercilious Cynthia. Cynthia, an eight-year-old adolescent, was one of the two original animals released by Barbara Harrisson on Bako Island in 1962 in the first attempt to rehabilitate captive orangs. Yet after six years she was still not sufficiently independent to live completely wild. As I watched her weaving sticks together to make a little ground-nest I wondered about the value of such schemes. Was it really worth all this effort when the apes seemed so reluctant to return to their natural home?

James, the game-ranger, beckoned me to follow and led the way across a rickety log bridge. We passed quite suddenly into another world; the dark cool realms of the Bornean forest.

A narrow path wound through the thick vegetation between the buttressed roots of huge trees. Scattered among their lofty crowns were twisted piles of branches where orang-utans had constructed sleeping nests. James heard something ahead and we crept forward slowly. He stopped and pointed. A large red orang swung down with graceful ease from a tangle of vines and stopped in a small tree to watch us. At the camp on the ground the orangs had seemed comical and ungainly but here among the trees the long powerful arms and russet hair looked just right. But most exciting of all, clinging to the fur on Joan's side was a tiny orange baby. On one of her trips into the forest Joan had mated with a wild male and the baby Joanne, now just a year old, had been born in the camp.

As we sat quietly watching them we heard a sound to the left. A smaller orang swung lazily through the trees towards Joan and her baby. The newcomer peered closely at the infant and gently put out a hand to tickle it while Joan looked benignly on. James explained that Coco always stayed close to Joan when she travelled in the jungle.

As I watched the two youngsters playing happily, I wondered how long it would be before I witnessed such sights in the wilds of the Ulu Segama. Already I could feel the difference in tempo between these easy-going gentle animals and the excitable black chimps I had known in Africa. I would have loved to stay longer in the Sepilok but had to hurry back to Sandakan if I was going to be ready to fly south the next day.

At Lahad Datu the Forest Department helped me to find a small boat and two native Dusuns willing to accompany me on the trip. I also learned that a geological survey team was due to set off into the Ulu Segama in two days' time. At the government rest house, beside the airstrip, I met Meng, the Chinese geologist in charge of the survey. He listened impassively to my plans and suggested that we should leave together.

'That way,' he said, 'at least I can see you safely as far as River Bole.'

There our ways would part as I wished to go up the tributary Bole while Meng would continue up the main river. He politely hinted that I was foolhardy to attempt the trip with only one boat, two men and no gun; also that three months was too long to stay in the jungle on one trip, but he made no attempt to dissuade me from my venture. For his own six-week expedition he was taking two long boats and a team of fifteen Iban Dyaks from Sarawak, all experienced jungle men. He had encouraging news, however, about the region. On a previous survey of River Bole the geologists had found several orang-utans and tracks of '*Badak*', the rare two-horned rhinoceros. There were many elephant in the area but these had caused no trouble. The men had had to shoot three honey-bears, however, and the Ibans regarded these bad-tempered creatures as the most dangerous animals in the forest. A few years earlier honey-bears had killed two men at a nearby timber camp, ripping open their bellies with sharp curved claws.

When I reached the broad muddy river the boats were ready. My two quiet Dusuns looked very tame next to Meng's muscular team. With their tattooed necks and shoulders, pierced ear lobes and long hair, the Ibans looked every man's picture of the wild men of Borneo. Beside the two long boats with their powerful outboard motors my own little *sampan* looked rather pathetic. It was already overloaded and, lacking a motor, would be very slow against the fast river. I decided to go ahead with Meng and arranged for my own men to meet me at the mouth of River Bole as soon as they could get there.

After transferring enough stores for a few days I climbed into the larger boat. The motors roared into life, I waved farewell to the Dusuns and the boats pulled out into deeper water. For several miles the thatched huts of *kampongs* lined the banks and smiling children splashed at the river's edge. Gradually the huts became more sparse and the vegetation wilder. Large rocks peeped menacingly from the water and the swirling currents slowed our progress.

We rounded another bend and the scenery was suddenly

Above Foster mother wins a kiss

Overleaf A Kampong orphan due for release

Above The bald clown, Jippo and, below, another youngster surprised at Sepilok

Overleaf Hills rose steeply from the riverbanks above the Ranum

transformed by the great white cliffs that rose on either side. A band of monkeys scampered over the rocks and swirling flights of swifts swooped in and out of the caves that pocked the cliff face. The Iban in front of me turned and grinned to show his blackened teeth. He pointed into the deep river pool and said, 'Banyak Buaya (many crocodiles), Tuan.' I nodded with uncertain understanding. We passed on into the uninhabited headwaters, dwarfed by the limestone walls and towering trees.

Hour after hour we worked our way in the fierce sunshine. We were constantly hitting submerged branches and rocks in the fast water so that the propellor pins had to be changed many times. At each set of rapids the Ibans lowered long poles into the river. The poles bent with the strain as they leaned and at every heave our boat leaped up and forward like some many-legged insect. When even this failed we waded through the powerful current, pulling the boats up with ropes.

The Ibans enthusiastically pointed out the troops of red, grey and brown monkeys that leaped among the trees as we passed. Eight-foot-long monitor lizards basked on sunbaked rocks and electric blue kingfishers flew by, shrieking. Snakebirds played hide and seek, disappearing for minutes under water then surfacing as we drew near and flying ahead to play again. Overhead a red-casqued hornbill glided in a clear sky, its jagged wings cutting the air with a loud whistle as it mocked us with a cackling cry. On either side the tall dark forest offered tempting shade among the great trees, plumed with epiphytic ferns and gently swaying lianas. Elsewhere the hills rose steeply from the river banks, covered in a dense tangle of prickly jungle.

The motors droned on and the Ibans chatted gaily. They were excited about the trip as it was four months since their last expedition. I was sunburnt and sore, my water-bottle already empty. I copied the Ibans and scooped handfuls of warm, muddy, river water to my mouth. The men pointed out a tall tree. I gathered that something was very good to eat but could see nothing. Only when we drew near did I realise that it was an enormous roost of giant bats. Thousands and thousands of these unholy beasts hung in the treetops like some rich crop of fruit. They fanned themselves gently with folded wings and started twittering at our approach. The Ibans shouted and beat the water with their paddles; the bats poured out of the tree till the sky was blackened with their wheeling, shrieking bodies and broad membranous wings.

A mile up-river, we stopped on a large tree-covered island. Within minutes polythene sheets were erected over a frame of saplings, a fire was lit and rice was on to boil. I wandered down to the river, where two of the Ibans were having a competition with catapults. I might not have impressed them with my boatcraft but stone-throwing was a game I did know. I picked up a smooth pebble and sent it flying some thirty yards beyond their own efforts. They watched with interest as I selected another stone and hurled it towards a flock of swifts gliding low over the water. The stone struck the leading bird, which dropped lightly into the river. I could not have hit another if I had stayed a week but to the Ibans this was a tremendous omen and they rushed back to tell their comrades of the feat. I was sorry about the swift but was greeted with new respect at the camp-fire.

Over supper of rice and tinned fish I asked Meng about the crocodiles at the limestone cliffs.

'Ah,' he said, 'that is a very old story. The place is called Tapadong. To-day the caves are only used for collecting edible swifts' nests. Long ago they were used as a burial ground by the river people. When men first came to the Segama, however, the Tapadong was the home of a Garuda (an enormous eagle) and Tarongari, the giant crocodile. The Garuda watched from its clifftop nest ready to swoop down on anyone venturing near, while Tarongari lay in the water where he could upset and devour any boats that came up the river. Between them these two guarded the river at the Tapadong so that no one could pass. One day the men of the river decided to kill these self-appointed guardians. They built a raft, baited it with a dead deer surrounded by sharp bamboo spears and pushed it towards Tapadong. From a great height the Garuda struck down at the craft but pierced itself on the spears and hastened back to its eyrie. The wounded bird decided to fly away to another land but in its struggles to get airborne dislodged two huge rocks, which tumbled down to crush Tarongari below. For weeks the water was foul with decay, but from that time on the river people have been free to use the caves. The two great rocks can still be seen at the foot of the cliff and the place is said to be haunted by crocodiles. I don't really think there are crocodiles so near the *kampongs* as they are keenly hunted for their valuable skins, but there are certainly some very large ones farther upstream.'

As we sat round the fire drinking coffee an eerie wail started up in the forest and was repeated from all directions. Every evening this unsettling sound heralds the dusk in the forest. The call is made by an animal called *Tanggil*, yet none of the Ibans had ever seen it or could tell me what it was. As darkness fell the wailing died down and a chorus of deep-croaking tree-frogs struck up in its place. In ones and twos the fruit bats of the afternoon came sailing overhead in their search for fruit and young leaves. They flapped with the slow beat of enormous birds and the mist and darkness exaggerated their five-foot wingspan to monstrous proportions.

Meng drew my attention to the hands of Gaun, the Iban headman. An Iban does not feel fulfilled as a man until he has taken a human head. For each man he kills he may tattoo one knuckle of his hand. Most of our companions had several decorated knuckles but Gaun's hands were coloured from wrist to fingertips. Like many of his race he had volunteered to scout for the British forces during the Malayan emergency, when he had killed many terrorists. Since British rule made head-hunting illegal, the Japanese War and Malayan Emergency were the only opportunities for the Ibans to revive their old traditions. Gaun explained that there were still a few communist terrorists in Sarawak whom Ibans could hunt without fear of Government reprisal, but that many youths had to be content with the head of an orang-utan. I asked him how one hunted orangs.

'An Iban,' he said proudly, 'can kill an orang with his *parang* (sword), but it is easier to use a blowpipe.' Being large and slow-moving, the orang makes an easy target but he is very strong and needs three or four poisonous darts before he vomits and falls. For a man a single dart is sufficient. He described how the men collected the poisonous sap of the *Ipoh* tree and mixed it with the juice of chilli peppers to make it more painful.

'It is important that it hurts,' he said, 'for when the animal rubs the sore he spreads the poison quickly.'

I left them at the fire and walked back to the shelter, where I tried to find comfort on a rush mat on the hard gravel. As I lay listening to the soft tones of their chatter, I hoped that my own head would not prove a temptation in the night. Despite the buzz of mosquitoes round my ears, I was soon asleep. I was vaguely aware of distant gunshots, but it could have been a dream.

I was awakened by a young Iban offering me a bowl of rice and stew. I peered at the latter suspiciously.

'Pelanduk (mousedeer),' he said.

Remembering the gunshots in the night, I nodded my thanks and tucked in with hearty appetite.

By sunrise we were already on our way and at midday reached the T-junction where Bole joins the Segama in a swirling whirlpool. The men drew up the boats on the gravel beach facing the tributary and Ibans hurried off in all directions. Within fifteen minutes they had made a floor of long strips of bark and erected my canvas above it on a wooden frame. They showed me which vines I could drink from and what stinging leaves to avoid, then bade their farewells and returned to the boats. Meng and his cocky headhunters rounded the bend in the river and the sound of motors faded in the distance. I turned away and walked back towards my tent. Now I was alone and it would be several days before my Dusuns could reach me. I gazed up at the tall trees around me. Perhaps it was only my imagination but I felt as though I was being watched, that an animated jungle was studying the pale stranger with unfriendly eyes.

Chapter two

A Hostile World

I knew that the Segama River ran east–west, so, setting a compass line for due north, I set off to explore my new surroundings. I climbed the steep bank behind the camp, passed through a narrow belt of dark forest, then descended into a thick tangle of secondary growth beside a narrow stream. I waded waist deep through the cold water and scrambled out on to the muddy bank. My way was barred by a prickly mass of creeper-covered bushes. Wielding my sharp *parang*, I began to slash a passage through the thicket, but progress was slow. As soon as I made a gap it was immediately filled by more vines and branches tumbling down, pushed by the weight of the creepers above. Eventually I succeeded in making a narrow tunnel and crawled through on my hands and knees.

I felt a sting on my rear and discovered a large slimy leech settled down for a feed. I had been expecting leeches and came equipped with a salt-soaked cloth, but I was revolted by the suddenness of this first attack on such a vulnerable part of my anatomy. Unwrapping the salt pad, I placed it over the loathsome creature, who instantly withdrew its mouthparts and dropped to the ground, writhing horribly and secreting a flood of slime in its attempts to wash off the salt. Two more leeches looped towards me across the dead leaves, waving their trunk-like bodies in my direction to scent the way. Taking a deep breath I plunged on past them. Clinging thorns tore at my legs and clothing, but I pushed on, head down, till I came again on to firmer ground where the forest floor was more open. I paused to examine the damage and to my horror found another three leeches climbing up my legs. I hastily plucked one off with my fingers and tried to throw it away, but each time I freed one sucker the leech clung on tenaciously with the other. When at last I managed to scrape it on to a tree-trunk, I chopped its waving body in half with my *parang*. The other offenders suffered a similar fate.

I continued north through the primary forest where little light could penetrate the thick canopy. In the gloom I completely failed to see the fine, barbed tendril of a *rotan* palm until it caught me across the face. Instinctively I pulled back but the barbs bit deeper and two more tendrils swung down to fasten on to my hat and back. Angrily I pulled out my *parang* and slashed to right and left but only tore the skin from my wrist and became more tightly entwined. Learning my lessons the hard way, I slowly freed myself, one tendril at a time, then set off again paying more attention to where I was going.

A steep hill rose before me, and by the time I reached the top my legs were aching, so I paused for a few minutes, seated on a rotting log. I surveyed the surrounding treetops for orang-utan nests. So far I had seen no sign of my quarry but as it took all my concentration to move through the forest without mishap I had hardly had time to glance up into the canopy. The hill led on to a narrow ridge and here the going was much better. Obviously I would have to get to know the terrain and learn the best routes if I was to travel through the jungle with any ease. From the ridge I could look into the treetops on either side, although the

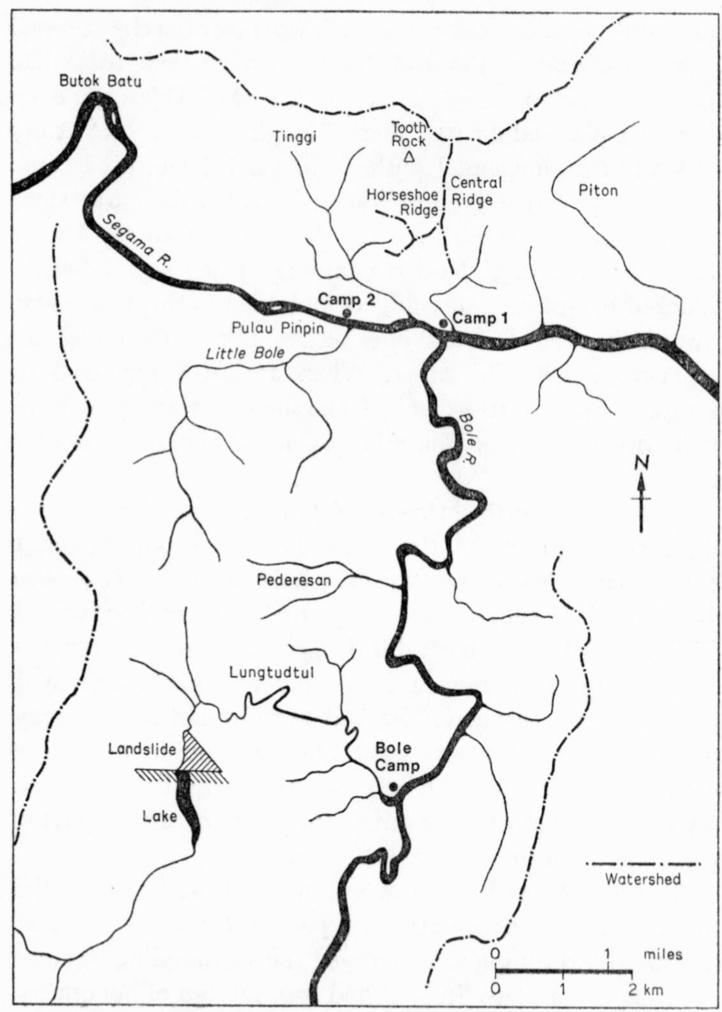

Segama study area, Sabah

solid wall of foliage obscured my view of the surrounding country. I tried to map some of the landmarks I had passed but had no idea how far I had come in my hour of travelling.

The ridge ended abruptly and I slid and scrambled down a sharp drop towards the bubbling stream below me. I clutched at a tree for support but hastily withdrew my hand, smarting from the sharp spines on the trunk. I relieved some of my hostile feelings towards the forest by slashing a raw, pink wound in the bark of a large *Seriah* tree to mark my route. The stream bed made a natural easy path through the encircling forest. I made my way carefully over the slippery rocks and the cold water seeped through my soft canvas shoes to soothe my sore feet. A large tree had fallen during a recent rainstorm and where it bridged the pool I left the stream, climbed the bank and headed north again.

A deep red stain on the top of my shoe drew my gaze. Inside a cluster of leeches were feasting on the side of my foot, bloated revoltingly from their meal. I hurriedly brought out the salt pack to remove them. Yet another clung to my leg and I rolled it mercilessly between my fingers until both suckers came free and I could flick it away. Blood oozed thickly from my foot and when I wiped the wound clean I could see the small Y-shaped holes where the leeches had bitten me. I had never expected them to be so plentiful or such determined travelling companions. They were obviously going to prove quite a problem as I could not afford to have open wounds. In the tropics the slightest scratch can easily become infected and as I looked at my torn arms, legs and face with alarm I wondered how much would be left of me in a few months' time.

A crashing in the branches attracted my attention and a small brown monkey leaped agilely through the treetops and hurried away down the hill. A few minutes later I found what I was looking for, the nest of a wild orang-utan, a pile of broken branches, sixty feet up against the trunk of the tree. It was black with age but a nest nevertheless, so that I knew that orang-utans must visit this part of the woods. It didn't look much, that untidy heap of leaves, but it did wonders for my fast-flagging optimism. Somehow I would do it. Somehow I would find these elusive beasts in their own awful terrain and learn their secrets.

The journey home took only half the time and I arrived

with great precision exactly where I had entered the forest just above my camp. I lit a fire to boil my rice then wandered down to the river to soak my tired, aching body in the cool, brown waters. Night comes early in the tropics and there was little to do when I had eaten. Although it was only seven o'clock I put on some dry clothes, rolled myself up in my blanket and went to sleep. Nothing was going to keep me awake to-night, not mosquitoes, nor the uncomfortable lie of the bark floor, nor the strange calls of the forest night.

I slept like I hadn't slept for a week and it was already ten o'clock when I woke again to cook myself breakfast. I swam in the river to ease the stiffness from my limbs, then had an easy morning laying out the stores I had brought and catching some of the fantastic, multi-coloured butterflies that flitted along the riverside. By midday I was ready again for any trials the jungle could offer.

Back in the forest I found a narrow path parallel to the river. Old elephant droppings left me in no doubt as to who was responsible for this open roadway but I was only too willing to take advantage of such a luxury. When the path petered out I followed a compass course north-west and was very relieved to find I did not have to cross the terrible stream that passed close behind my camp. I cut diagonally across several small ridges and intersecting streams but was making good progress when I arrived at the bank of a small river. My map showed this to be a tributary of the Segama and gave me some idea of how far I had come. I waded across a shallow stretch to a sandy beach where there were many footprints of wild pigs and a small cat that I thought could be the clouded leopard. I followed along the course of the river, but it meandered in such wide arcs that I did better when I left it and followed my compass again.

Branches swayed and I was suddenly startled by the loud 'cha-chack' of alarm as a group of bright red monkeys scattered away noisily through the treetops, then stopped to hide as soon as they were out of sight. I was amazed at their vivid colouring, even brighter than the orang's, and wondered what advantage it could have in this leafy green environment. I continued up the gentle incline, stopping every few minutes to listen for animals and to pick off the ever-persistent leeches.

Loud crashings through the undergrowth marked the flight of two pigs which I had disturbed whilst they were cooling themselves at a mud wallow. A pool of water had gathered in the hollow left by an uprooted tree. Regular use by pigs had turned the pool into a bath of liquid, yellow mud and they had undermined one side to form an overhang. I could see the imprint of their hair where they had been lying. Near the top of the hill I heard soft hoots from a party of gibbons. I crept closer and hid behind a convenient tree to watch their approach, but saw them only briefly as they flitted past like grey ghosts in the branches. Their hooting grew fainter as they moved away down the slope I had just climbed.

I followed the ridge round where it curved to form a horseshoe. I had nearly reached the end and was about to turn back when I sensed, rather than heard, a large animal moving about in the trees ahead. I knew even before I saw its shaggy, red shape, that I had at last found my first wild orang-utan. Tingling with excitement, I moved slowly forward, making no attempt to conceal myself. It was a large male with a magnificent long coat and broad face. I christened him William the First and took out my binoculars to see him better. He did not seem to notice me but continued feeding, slowly munching on the large green fruit he had found. William moved deeper into the branches out of my view but just then I noticed another orang-utan, a large female. As she climbed up into the fruit tree I saw that she was carrying a small baby about the same size as Joan's at Sepilok. Mary, as I called her, started feeding on the green fruit, but although she did not look at me I was sure she was aware of my presence for she started to make a loud gulping noise and shaking branches, threatening. William also took up the call but preceded his gulps with a curious sucking squeak. He moved across into another tree, bending branches perilously to bridge the gap, then disappeared into a thick tangle of climbing bamboos. Loud crunching noises suggested that he had not abandoned his meal. Mary continued to feed, still avoiding my gaze, but after another ten minutes she made a dignified retreat, moving slowly and carefully to the bamboos, her baby, David, still clinging tightly to her side.

I tried to go round the bamboos from the other side, but by the time I had forced a way through the thicket there was no sound or movement from the orangs. I waited, quietly scanning the surrounding forest, then saw Mary high up in a tree some fifty yards away. David was playing alone, dangling at the end of a springy branch by one arm, while his mother kept a watchful eye on him. He gazed blankly in my direction as he swung, then suddenly he saw me, ducked his head and, giving shrill whimpers, hurried back to his mother. Mary gathered him protectively in her arms and, with David clinging tightly to her side, she climbed along the branch, which bent under her weight, and swung across into the trees beyond. It was getting late, and since the orangs would nest nearby I made no further attempt to follow them. I headed home, cutting saplings to mark my route for the morrow.

As darkness fell the forest was filled with the eerie echoing call of the *Tanggil.* I crashed blindly on through the undergrowth, regardless of thorny creepers or anything but the glowing face of my luminous compass. The moon shone brightly, leaving scattered patches of light on the forest floor. I kept travelling south-east and an hour later was back at camp taking my evening bath. There was still no sign of my men, so I cooked my rice, opened a welcome tin of meat and made ready for bed. Now that I had time to review my encounter with the orangs I was much too excited to sleep. I tossed to and fro on my hard bark couch and regretted not having a net to keep out the plague of sand-flies and mosquitoes.

Long before daylight I was up packing tins and a plastic cape into my bag prior to hurrying back into the forest. I followed the elephant path to the little river, then headed north-west again by compass. By nine o'clock I knew I had missed the horseshoe ridge and must be far beyond the area of the previous day. I headed downhill back towards the little river but when I reached it found no more than a stream. Certainly I had come too far. I splashed along the stream bed, following each twist and turn. Then I noticed footprints in the sand – footprints of a large animal with three toes. Only one animal could have been responsible – *Badak*, the two-horned rhinoceros. The tracks led down-stream and looked very fresh. At any other time I would have followed them in the hope of seeing this rare, almost legendary, animal but just then I was desperately trying to get the study off to a good start, and it was orangs I had come to see, not rhinos. I was not to know that it would be over a year before I saw rhino tracks again.

Leaving the stream, I climbed a small hill and followed an animal path on to another ridge. I noticed a movement to my left and dived for cover. A tiny brown *mawas* climbed quickly down from the top of an enormous tree and swung gaily through the branches towards me. As it drew near I saw it was larger than I had first thought, probably about three years old. He was a delicious, dark chocolate colour, with pink rings round his button eyes and a shiny bald head. I named him Midge. He climbed up on to a branch only twenty yards away but did not see me squatting behind a ground palm. There was a heavy, crashing sound and a small tree swung in a wide arc towards him then quickly away again as a large orang-utan transferred herself to Midge's tree, then climbed up to join him. For a moment I wondered if these could be the same animals I had seen the previous day, but the juvenile was too big and I thought we were still some distance from the horseshoe ridge.

As I wanted the animals to get used to me and lose their fear, I rose from my cover, moved casually towards them and sat scratching at a small hole in the ground in which I pretended a keen interest. The animals did not seem particularly disturbed at my sudden appearance but both began calling – curious sucking squeaks, accompanied by deep guttural grunts from the female, Margaret. She watched me for a few minutes, then climbed into a taller tree and began nibbling at the young leaves, bending the twigs to her mouth with her free hand. Tempted by these fine delicacies, Midge joined her. Every few minutes Margaret left her meal to watch my progress at the hole, uttering her strange vocalisations and shaking a branch or two before returning to feed. Becoming increasingly annoyed at my unwanted company, she stopped eating to give a violent display. With a twist of her arm she sheared a small branch clean from the tree and dropped it with a casual gesture, leaning out to watch it tumble noisily down. Vocalising with a loud series of croaks, 'lork, lork, lork', and chomping noisily she broke off another branch and swung it threateningly backwards and forwards several times. She dropped this also, then looked up anxiously to see how I was reacting to this show of strength. Not wanting to disappoint her I pretended to be very impressed and crept quietly away to sit and watch from a greater distance.

A small female orang had been hidden in a nest overhead, however, and she, too, started shaking branches so that I had to jump aside to avoid a barrage of broken twigs. I moved back but, undeterred, the adolescent climbed overhead to drop more missiles. A heavy piece of dead wood crashed to the ground, narrowly missing me, and I made a more serious retreat to a vantage point from where I could watch all three animals. The new female quietened down and sat watching me. I decided to call her Milly. Margaret and Midge were still feeding on leaves and I had time to take a good look at Milly through my binoculars. She looked about seven years old, had dark chocolate fur and a narrow black face with close-set eyes and broad muzzle. Her hair was so sparse that she looked like a large spider as she moved from branch to branch.

Margaret and Midge started to travel down the slope, with the characteristic slow climbing and tree swaying that enables orangs to move about quite independently of the ground. I left Milly watching intently from her tree and followed after her departing friends, keeping at a fair distance so as not to disturb them. They climbed into a tall tree bearing small, egg-shaped *rambutan* fruits. Midge perched out of sight at the very top of the tree but Margaret started feeding lower down. She gave occasional deep grunts but seemed resigned to my presence and ate solidly. A large animal, lighter in colour than her companions, she had piercing, bright eyes and a scruffy red moustache. The hair on her head came forward in a tight mop framing her broad face.

Having cleared all the fruit within her reach, Margaret shifted her position and started on another branch. Pieces of fruit shells dropping from the higher branches showed that the invisible Midge was feeding also but there were no sounds from up the hill where we had left Milly. Gradually their feasting became slower and Margaret spent long periods sitting quietly, supporting herself by holding on to the branch above with one long arm.

The day warmed up fast as the sun approached its zenith. Margaret moved lazily to the middle of the tree where she bent two branches across the crutch of a larger bough. Satisfied with her handiwork she lay down on this crude couch. Midge was still active, so I moved a few yards nearer to get a better view. I could see him playing, bending and shaking branches. Having pulled several together around himself, he tucked them beneath his feet to make a little play-nest. He piled more branches on top of his head and thumped them up and down across his back in play, finally outspreading his arms to let them fall around him. Bored with his games, he left the nest and climbed down to Margaret. With one hand on her back and the other clinging to an overhanging branch, he sat resting, watching me in silence.

It was nearly two hours before they stirred again. They seemed to have forgotten me and had a quick snack before moving back up the slope, crossing from one tree to the next with practised ease. They returned to the ridge only fifty yards from where I had first seen them and headed knowingly towards a tall, reddish-barked tree. Midge first and Margaret close behind, they climbed up the tortuous stem of a thick liana that curled round the trunk and clambered out along a large branch. Prickly, yellow fruit, the size of oranges, hung in clusters and they started to feed. Several chewed pieces on the ground suggested that the orangs had been here before. Margaret had no difficulty in breaking the tough stalks and biting through the thick, spiny shells to remove the fleshy contents but poor Midge had more of a struggle. He bit and tugged furiously with both hands to rip the fruits free and spent several minutes opening each when he finally detached it. Sometimes he had to hang completely upside down, suspended by his hook-like feet, to tackle a fruit below him. While the orangs were busy above, I crept beneath the tree to select some dropped pieces of fruit to add to my collection for identification. The fruits were as prickly as a hedgehog and oozed white jelly. I cut one open with a knife and sampled the large bean-like seeds, which were sweet and crunchy like a brazil nut kernel.

A large hornbill landed overhead, his pinions whistling in the wind. He raised his heavy, helmeted beak to give a series of single sad calls. His hoots continued for about three minutes, becoming gradually faster until they raced to a staccato climax which broke into a hideous, chuckling

Supercilious Cynthia

Opposite Just blowing bubbles and, below, a wide-eyed orphan

Above Jippo and friend stride out
Below Reclining like a golden Buddah

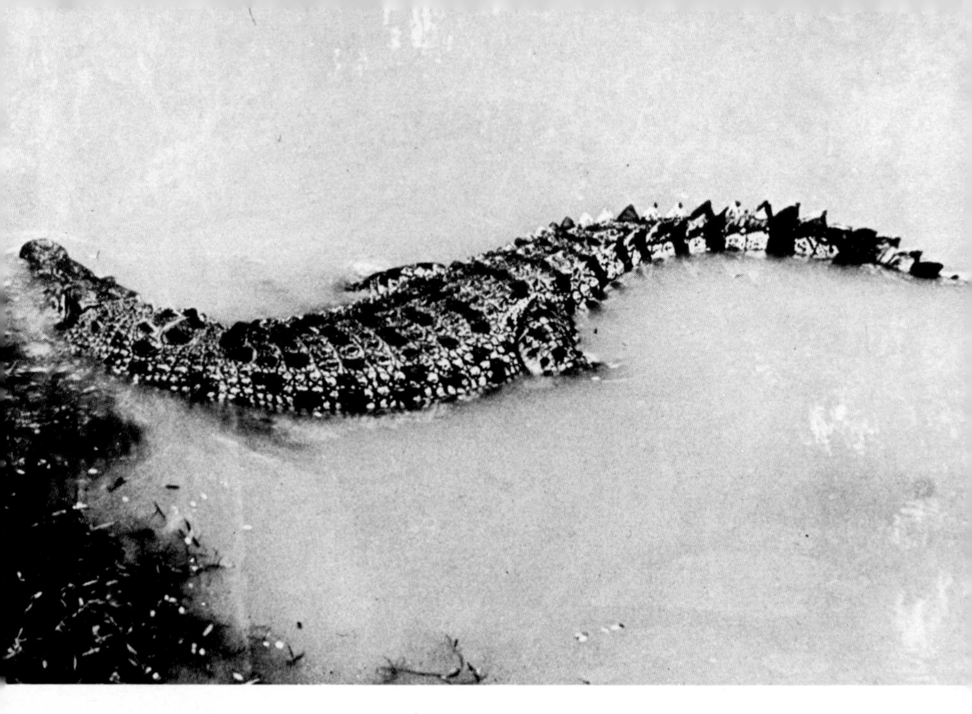

There were several large crocodiles in the Segama

The rivers became swollen with winter rain

cackle. A few minutes later, his song over, the bird flew off again, his long tail streaming behind. Unperturbed, Margaret fed on but Midge came down to sit by his mother again. He tried to take a piece of fruit from her hand, but she pulled it sharply away. Eventually, after feeding for over two hours, she lay down on a large branch and went to sleep. Midge, bored by such slothfulness, played happily by himself, swinging and dangling from the branches or breaking them off and piling them over his head as he had done earlier in the day.

Towards five o'clock Margaret decided it was time to move and clambered down the thick liana. As they crossed through the trees towards me, I had no time to hide and they were only ten yards away when they saw me. Margaret started her gulping noises and climbed determinedly past me but Midge was more anxious and whimpered quietly as he made a wide detour to overtake his mother. He tried to ride on her but she pushed him firmly away and he raised his shrill little voice in another whimper. I kept my distance and they slowed down to move steadily along the side of the ridge. After twenty minutes they stopped out of sight. I came closer and waited. Two black pheasants bedecked with broad, white tails came running along the ground. They turned to face each other, beaks raised, and chuck-chucked angrily before one turned tail and was again chased by the victor. They vanished as quickly as they had come but there was still no move from the orangs.

As darkness fell the *Tanggil* started his mechanical wailing and I felt a chill as I realised it was now far too late to get back to camp and I would have to spend the night in the forest. A crashing sound ahead told me the orangs were on the move again. They were farther down the slope than I had thought and I hurried to catch up, but, by the time I had got fifty yards, all was quiet again. Above me was a large green nest that must have been used recently, perhaps the night before, and I felt sure that my animals must be nesting close by. It was the end of a typical orang day – a day of resting and feeding and leisurely travel from one fruit tree to the next.

With my friends safely abed, I set about finding a place where I could spend the night. A mass of leaves had collected between the two broad buttresses of a large tree which would give me ample shelter. After some searching I found a long piece of spiny liana and fixed it around the open side of my eyrie to keep out any wandering pigs or other unwanted visitors in the night. Some cut saplings strengthened the palisade and the leaves made a comfortable bed. I was in great need of a drink but had no water-bottle, so after eating a meal of cold meat, I took the empty tin and set off in search of water, guided by the wavering light of my pen-torch. There was a steep drop behind the shoulder of the hill and I slithered down to the small stream that had cut this gully. Thirstily I drank a tinful of the clear, sweet water then refilled the makeshift vessel to carry it back up the slope. I slipped, grabbed a bush to save myself but lost half the water. Climbing more carefully with the precious remainder, I found my tree without further trouble.

It was surprisingly comfortable in my nest, far more so than the hard bark couch at camp. I slipped a pair of trousers over my shorts and donned a long-sleeved rainproof jacket. Pulling my plastic raincoat over myself, for it was already getting quite cold, I tried to go to sleep. Darkness closed in blacker than anything I had known. The trees blotted out the sky and any friendly stars. As my eyes became accustomed to the dark I was surprised to see glowing points in the surrounding night. At first I thought my luminous compass must be responsible but found it safely tucked away in my bag. I shone my pen-torch at a faint glow near my face but found only dead twigs and leaves. When I extinguished the torch I again saw the white glow from the forest litter. I picked up a glowing leaf and examined it closely. It was streaked with a lacy web of fine, luminous lines. Just outside my sanctuary a cluster of tiny mushrooms glowed like little yellow beacons. Elsewhere on the ground and tree-trunks tiny points of green and blue shone like the lights of a great city seen from an aeroplane high in the night. One of the largest spots was moving and, on closer inspection, proved to be a crawling grub, the row of glowing spots on its sides giving it the appearance of a miniature train chugging slowly through the night. More flashes punctuated the dark as fireflies flew through the treetops, winking their messages of love to anyone who could interpret the signal. I wondered why I had never read about this beautiful array of forest night lights and realised that probably few people had ever seen them, for those who sleep out at night raised off the ground and blinded by a glowing camp-fire would never see such sights.

The tree-frogs bleeped their songs amid the chorus of the jungle night. From down the slope came grunts from my nesting orangs and, far away, a strange groaning like an old man in great pain. Sleep came in broken patches amid curious dreams and spells of wakefulness. I thought I heard some animal moving about nearby but felt safe enough behind my wall of thorns. Enormous ants clattered through the leaves but none came to trouble me. Some time in the night it began to rain so, pulling the plastic raincoat over my head, I curled up tight against the sheltering tree-trunk.

I woke cold and stiff. A faint grey light filtered through the mist between the ghostly trunks of giant trees. Somewhere in the distance a single gibbon was singing to the whole forest to announce another day. It was five-thirty and I was ravenously hungry. As I had only one tin of meat left I ate slowly, relishing each bite, and drank the remains of my water. I buried the tins and packed my bag. Still wearing my trousers and jacket for protection against the bitter cold and damp which had seeped through every layer, I climbed down the hill to where the orangs had nested and sat to wait.

Branches swayed in the tree almost overhead and through the morning mist I could just make out the shape of a small, brown orang as she slipped away. I hurried after her as she swung quickly from tree to tree. At last she stopped to rest, there were slight movements, then all was still. She could not have left the tree without my seeing her so I stayed, watching for any sign of activity. It was an hour before she again slipped quietly away and hurried through the treetops. She climbed into a dead tree, shrouded thickly in a mass of creepers, and disappeared into the green tangle.

Slowly the mists cleared, the birds started to sing and the jungle came to life. Families of gibbons chorused back and forth over their jealously-guarded boundaries. Flocks of squawking pied-hornbills chased one another noisily through the treetops and the sawing of cicadas rose in volume as day began to warm. I removed my jacket and trousers and sat steaming dry in a patch of morning sunshine waiting for my orang to move again. I had only caught a brief glimpse of her in the morning mist but was sure it must be Milly and that the family had reunited at the nesting site in the evening. For two hours I sat waiting. Had I really seen her climb into that tree? Could she have slipped away unnoticed? I became more restless and uncertain. I leaped up and hurried back to where I had last seen Margaret and Midge. I searched the whole area. There were several green nests where they might have spent the night but the place was now deserted. I returned to Milly's tree but there was no sound. Perhaps she was still there hiding or perhaps she had slipped off while I was away. I decided not to wait any longer and climbed back up the ridge to head for home. The slanting rays of the sun reflected off the wet, shining leaves made dazzling patterns, adding a beauty to the jungle I had not noticed before. I was soon drawn back to more mundane thoughts, however, for the morning damp had brought the leeches out in their dozens and I felt a cold slap as each lodged on to my legs from the low foliage as I passed.

I was happy with the way things were going. I had found my red apes and already collected several hours of information on their travel, feeding and grouping. I was concerned at their shyness but hoped they would soon become accustomed to my presence so that I would be able to watch their undisturbed behaviour. Moreover, I had proved that I could sleep out with my subjects at night and join them again when they left their nests in the morning. Provided I could find ways of maintaining myself in the jungle I should be able to stay with one group of animals for several days at a stretch.

I bumped into two families of leaf-eating monkeys. They were an incredible sight with their black faces and bright orange hair which rose to a point in the middle of the head, like the hats of the workers in the *padi* fields. They rushed off with loud alarm. By midday I was back on the elephant path following the Segama down to camp.

Chapter three

The Solitary Ape

My two men had at last arrived with the rest of the stores and had replaced my tent with a compact stilted house. They had cut long poles in the forest and bound broad strips of bark with vines to construct a sturdy platform some three feet above the ground. A wide ladder led up to the hut, which was roofed with a large canvas. At one end was a separate kitchen unit of bark and thatch equipped with our curious collection of blackened cauldrons. Cooking preparations were already under way and a fish stew bubbled in a large, round-bottomed pan over the smoky fire.

My companions greeted me with relief for they had been anxious when I did not arrive the previous night. In broken Malay I tried to join in their cheerful chatter. Likad, the younger of the two, proudly showed me the large pile of catfish they had caught. Tulong, some ten years his senior, and of powerful stocky build, pointed to a small fig tree on the other side of the river. The branches were thickly laden with round, red fruit and even from that distance I could see the water swirling beneath, where the fish fought over the sweet fallen figs.

We enjoyed a delicious feast of the fatty fish and a pot of

greens prepared from the young tips of ferns that grew along the river banks. It was delightful to warm myself and dry my clothes at the fire. We talked endlessly. They wanted to know where I had been and whether I had found orang-utans. Halting frequently to look up words in my dictionary, I told them of my adventures. They seemed to understand and nodded agreement encouragingly. Likad interrupted to tell me that the small river I mentioned was known as Tinggi and that it was a good area for durian fruit, the prickly fruits that had so delighted my orangs. The specimens I had collected were still very young but when they are fully ripe durians are delicious and many Dusuns would come into the forest to search for them. Likad explained that orang-utans are far from popular among the natives because they eat the fruit unripe and thereby spoil the later crop.

In the evening we set out on a fishing expedition. The little boat wobbled violently at my every move until I learned to relax and sway gently with its motion. We paddled up-river past shallow rapids and into the clear waters of the small hidden tributary of Little Bole. Using long bamboo rods and short, hooked lines baited with a fig, the men cast to either side and caught a dozen good-sized fish. Tulong lent me his rod and I tried my luck. I was not very good but managed to land a couple of sprats. We drifted lazily back to camp, where Tulong pulled the canoe up on to the beach and made it fast.

Tulong was keen to join me in exploring our new surroundings and was a mine of information, reeling off the Dusun names for all the birds, animals and trees that we encountered. He was full of energy and spent several minutes pulling down long thorny *rotan* canes that climbed high into the trees. After scraping the springy creepers clean he wound them into a huge coil for carrying home. All the while Tulong chatted gaily, smoked an endless succession of cigarettes rolled in palm leaves and bounded about, hacking trees and cutting saplings to mark our path. We wandered over a large area including Horseshoe Ridge, where I had seen my first orangs. Tulong was very impressed by my compass and was quite amazed that I could make long circular trips through jungle I had never seen before. Unless he knows an area well a Dusun leaves a trail of bent twigs so that he can retrace his route back through the forest. I tried to explain the 'magic' of my little dial but my Malay and Tulong's knowledge of physics rendered the task hopeless.

With all this crashing about it was not surprising that we found no orang-utans, but as we were heading for home a long series of deep groans suddenly started up in the forest nearby, then tailed off into a string of bubbly sighs. I recognised the sound as that I had heard in the distance on my night in the jungle. Tulong assured me that it was an old male orang-utan calling. As I had never seen any reference to such a noise in the many descriptions of orangs I had read, I was very excited at this news. I took a compass-bearing on the call and we tried to find the animal but to no avail. We heard nothing further and after searching in vain for over half an hour abandoned our quest.

Every evening Likad and Tulong went off in search of food while I lazed about camp writing up notes and straining over my Malay vocabulary. When, an hour after nightfall, they had still not returned I was contemplating eating alone until I saw a torch flash down by the river. A few minutes later Tulong arrived, out of breath and very excited because they had killed a pig. They had stumbled across it sleeping in a deep wallow and while Likad held the torch Tulong had speared the dazzled animal. The paraffin lamp cast a pool of light where Tulong worked happily gutting the unfortunate beast. Soon pieces of wild pig were sizzling and charring in the flames. We toasted choice slices of fatty skin on sharpened sticks and ate them sprinkled with salt. Certainly the meat was very tasty though rather tougher than my weak jaws were used to.

Breakfast was again centred round the fire on the beach. There was enough meat for several days. Suddenly, as we sat eating, the same tremendous roaring groans as we had heard the previous day started up just behind our camp. I rushed off to locate the noise but the calls stopped before I reached the forest. About a hundred yards in I found a large green nest and on the ground nearby the shells of freshly opened *rambutans* and wild jackfruits. Tulong and I worked back and forth through the surrounding jungle in a systematic search but could find no sign of the caller. It was Likad who found the animal, sleeping in the crutch of a large tree, no more than fifty yards from our house. He was still there when Tulong and I arrived back in camp exhausted by our fruitless efforts. Harold was a large, dark-purplish male with a bare back. He sported swollen flanges on both cheeks and these made him look very fierce and ugly. Now I was sure that the long calls were indeed made by orang-utans. He sat resting for over an hour so that I was able to take several photos of him despite the poor light. I was soon bored by his inactivity, however, and left Likad and Tulong to keep an eye on him while I went off in search of my other orangs farther in the forest.

I had an enjoyable morning; the weather was fine and the leeches less abundant. I was beginning to get used to the jungle and managed to avoid most of the treacherous *rotan* tendrils. A party of pigs blundered into me, snorted and hurried away in single file. Of orangs, however, there was no sign. At midday I returned to camp hoping to continue observing Harold, but found that the men had lost him. Likad showed me where he had climbed to the ground and headed towards the terrible thicket and stream behind camp. It seemed an unlikely route for an orang to choose but there was no sign of him anywhere else.

It poured with rain and we all stayed in the welcome

shelter of our hut. The storm continued into the night and when I awoke in the morning I saw to my horror that, far from being camped about a hundred yards from the river, we were now only feet away and the boat was floating practically beneath the house. The gravel beach was a swirling, angry sea of leaping, brown waves and tossing, drifting timber. Of the fire and our glorious pig there was no sign and it looked as if our new house would soon be submerged, too, if the river kept rising. Leaving these domestic problems to the men, I began my search for Harold.

The big male proved difficult to find. I heard him calling several times during the next few days. He was slowly making his way up Tinggi stream, but whenever I took a compass-bearing on his calls and headed after him, he eluded me. It was four days before I finally tracked him down to a large fruit tree. Harold saw me at once but showed no fear. He was busily feeding on a type of wild lychee called *mata kuching* (cat's eyes). The ground below him was littered with several hundred discarded shells and stones, so he had obviously been there for some time. I noticed that he held his fingers in a curious manner but it was only when he at last climbed out of the tree that I realised they were completely stiff. The index finger of his left hand and two fingers of his right stuck straight out when the others were flexed, but this did not seem to impair his movement.

When I saw he was travelling directly towards me I became rather anxious. A thick liana passed over my head, bridging the gap between two adjacent trees and along this tightrope he now came. Swinging arm over arm he moved powerfully forward in a way that any gymnast would envy. It was too late to get out of his way so I crouched low and stayed quite still. His hanging legs can have been no more than a yard above my head as he passed, but he disdained to notice and carried straight on beyond me. For the remainder of the evening he fed in another *mata kuching* then sat resting quietly up the hill. There was a loud crash as a tree toppled over in the distance. Indignant, Harold raised himself high, facing the noise, and ballooned out his neck pouch with a loud, deep bubbling sound. Gradually this built up into the tremendous roars that I had heard so often. He reached a bellowing climax then gradually the groans subsided to low mumblings again. It was a great thrill to watch this magnificent display but I had to leave him to find a hideout for the night.

I climbed up the hill towards Horseshoe Ridge. The only level ground was a bare patch cleared by an argus pheasant as a dancing ring, and here I bedded down. In the valley Harold gave another long call while I covered my face with repellent to keep the mosquitoes at bay.

I was awakened by heavy crashing sounds from down by the stream. Only elephant could make such a noise and I suddenly felt very vulnerable. I was right in the open on an obvious pathway for any animal that should come up the hill. Hurriedly I collected my kit and, with the aid of my small pen-torch, moved off in search of a safer sleeping site. The side of the hill seemed far too steep for an elephant to climb so I settled there, by a small tree. Since I was in danger of rolling down the sharp slope in my sleep I arranged myself with one leg either side of the tree and lay down with my head pointing uphill. Almost asleep I felt a nudge at my foot. I kicked out to strike a frighteningly large resistance. An animal gave a tremendous snort and shuffled off with an incredible rattling noise. I groped frantically for my torch and in its dim beam saw a large porcupine snorting noisily only a few yards away. Bristling angrily, the beast plodded towards me. He was obviously not going to detour from his traditional path just because some fool was lying in the way. Unwilling to argue the point further I jumped aside and he trundled past me.

Once more I tried to sleep but the steep hill where I

thought no animal could possibly balance was fast turning into a major thoroughfare. A tremendous crash below announced that this was exactly where my elephant wanted to come. Hastily I grabbed my belongings and moved on again. I paused at the pheasant ring where I had been earlier but not for long, as the elephant was still advancing steadily. My strange smell on the night air had aroused his curiosity; he was following at a slow but relentless pace. Panicking, I rushed off into the night, heading for the only safe place that I knew, the buttressed eyrie where I had spent my first night in the forest. I was afraid to use my torch with the elephant so close but fortunately knew the path quite well, or so I thought until I became caught up in a mass of *rotan* tendrils. This was no time for careful extraction so, with a quick flash of the torch to see my way, I lunged forward. I painfully tore myself free and hurried on across the narrow ridge and up the hill to my sanctuary.

At last I found the tree, clambered round the buttress roots and collapsed with relief on my bed of leaves. I was safe at last for, even if the elephant should follow me here, a strong liana provided an easy escape ladder up the tree. Thankfully I heard the beast crash past some way down the hill but, nevertheless, I got little further sleep that night. When morning broke I set off straight for home. There in the pheasant ring was a fresh pile of large droppings. Imagining elephants behind every tree, I broke all records for the distance and it was several days before I dared sleep out in the forest again.

For the first month all went well. Gradually my knowledge of the forest grew and its geography began to fit together like a huge jigsaw puzzle. I was able to draw a map of the main landmarks and streams in the three square miles of my research area. Leeches, flies, mud, *rotan* and falling branches continued to hamper my work but I was getting used to such things. I maintained a Spartan camp – no bed, no books, no radio and no alcohol or coffee – in the hope that this would encourage me to stay in the forest as long as possible. My stock of tinned food was also strictly for eating away from camp. Only in the depths of the jungle did I allow myself to relish the forgotten flavours of baked beans, Spam and tinned milk. Tulong obtained some sheets of strong polythene from visiting Dusuns and these proved invaluable on overnight trips. With a sheet on the ground beneath me and another as a roof above I was both warm and dry when I woke in the morning.

Slowly my notebook filled with jottings on orang-utans. I had seen more than twenty different animals in some eighty hours of observation. After finding Harold again I was able to plot his daily progress through the forest from his calls. I also re-encountered Margaret and Midge and managed to stay with them for three revealing days before losing them in a rainstorm one evening. By now I could easily recognise my red friends and, as I delved deeper into their private lives, I began to realise what unsociable animals they were. Not for them the rough and tumble of community life. Orangs lead a withdrawn existence, quietly minding their own business and getting on with important matters like feeding and nesting. Occasionally two or three animals joined up to harvest a rich crop of fruit but they always returned to their solitary life without any sign of regret. The adults obviously preferred this hermit-like existence but I felt sorry for the infants with only tired old mum for company.

A severe attack of fever laid me low for three days but seemed to respond to my chloroquin tablets. I was soon up and about again but a week later the fever returned. Anxious that illness should not ruin the trip, I forced myself to go into the forest. With a temperature of 103° F., drugged with chloroquin and Disprin, I ventured as far as Horseshoe Ridge and returned by a different route. So long as I stared fixedly at my feet I could manage perfectly well, but if I tried to gaze up into the canopy my vision blurred and waves of dizziness swept over me.

I was nearly home when I saw a terrifying spectacle. For a moment I thought it was a trick of my vision. A huge, black orang-utan was walking along the path towards me. I had never seen such a large animal even in a zoo. He must have weighed every bit of three hundred pounds. Hoping he had not noticed me, I dived behind a large tree. I was in no state to defend myself, or run from him should he come for me, and I could recall clearly the natives' terrible stories about old, ground-living orangs. I held my breath as the monster passed within a few feet of me and let him get about forty yards ahead before I followed in pursuit. He was enormous, as black as a gorilla but with his back almost bare of hair; Ivan the Terrible was the only name I could think of. In my weakened state I was unable to keep up with his purposeful stride. I blundered on, nauseated, dizzy, and failing to see that he had stopped almost crashed into the great beast as he sat munching shoots. I realised the danger just in time and checked my pace. He ambled on again and climbed into a *mata kuching* tree laden with fruit. Ivan was far too heavy to venture out on to the frail branches, so settled comfortably in the middle of the tree and one by one broke the boughs in towards himself. Within half an hour the tree was stripped; its branches hung limp and torn. Ivan carefully lowered his great bulk to the ground; the days when he could swing easily through the canopy were long past and because of his huge size he was now forced to travel on the forest floor. He set off again at a spanking pace and, although I was keen to see whether such a large animal would still nest in a tree at night, I was too weak to keep up and soon lost all sign of his passing. Dazed and exhausted, I struggled home.

That night the deliriums of my drugs and fever combined to haunt me with terrible dreams of fiendish creatures taunting my sanity. I lit a candle and sat staring at its flickering flame until the safety of daylight rescued me from my nightmares. If things did not improve, I decided, I would have to return to town to see a doctor.

Luckily I had a wonderful day in the forest finding another calling male and a delightful mother and tiny baby. Both were deep chocolate colour and I christened them Nanny and Sambo. Sambo was the smallest baby I had seen and Nanny a loving and attentive mother. After a good night's sleep I woke to find my fever had passed, my head was clear and I was ravenously hungry.

Several times the men ferried me across the Segama so that I could watch the orang-utans who lived beyond the river. Here I met a courting couple; Bill, a handsome young sub-adult with long orange hair, and Jay. The happy pair spent a lot of time playing together and always nested close at night. Jay was small, trim and dark brown in colour but Mark, her juvenile son, had a blazing orange coat suspiciously like Bill's. Mark was excluded from his elders' play and wandered off alone but never out of sight of his mother.

Although my Malay was improving daily it was far from perfect and this led to a strange adventure. My men had dropped me about a mile up the Bole River with instructions to meet me again in three hours' time but they misunderstood and thought I meant three o'clock the next day. I waited at the rendezvous till dark, but as I had no food or bedding with me I did not want to stay out all night. I hid my kit beneath a log and plunged into the river to swim back to camp. Halfway home I heard a loud splash near the bank. Thoughts of crocodiles raced through my mind and I scrambled into the overhanging branches of a tree. I clung there, shivering in the inky blackness, my ears straining for the slightest noise but all was still. Cautiously I lowered my trembling limbs back into the murky water and swam on. At last I caught sight of our lamp on the distant bank and could hear the whirlpool where the two rivers met. Not wishing to risk this hazard at night I decided to cross farther down. I pushed off, straining every muscle, but did not seem to be getting any nearer to the far shore. I was swept round a bend by the strong currents and the friendly light was suddenly blotted from view. Here the river flowed less fiercely and gradually the dark line of trees grew closer. My foot hit a rock. I stood, stumbled, and stood again, then clambered out on to the sandy beach.

Back in camp Tulong and Likad were horrified that I had swum home in the dark and warned me of the dangers of such foolhardiness. Behind their scolding, however, I felt they were rather impressed by the feat. Highly-exaggerated accounts of the adventures of their '*tuan berhani*' (fearless master) were soon circulating around the Segama *kampongs*.

Towards the middle of August orangs seemed to become scarce. We had had little rain and the forest was drying up. There had been a good deal of leaf fall and little fruit was visible. I saw only one shy adolescent in a whole week and not a single new nest. My orang population appeared to have moved out of the area and I presumed they must have found valleys farther in the hills where food was still plentiful. I decided that we, too, should try new territory and made preparations to transport the whole camp a few miles up Bole River. Before our departure a boat arrived bearing Likad's brother, Sipi, and his three hunting dogs. Likad wished to return to the *kampong* to supervise the planting of his rice but Sipi would take his place on the trip. Likad planned to return in a month's time.

Tulong had been making a new boat down by the river. It was almost finished but not yet sufficiently stable to take much weight so he paddled this alone. Sipi, the three dogs and I travelled in my own boat but we were so heavily overloaded that progress was painfully slow up the fast-flowing Bole. At each set of shallows we had to wade in the water pulling the heavy craft along behind us. During these breaks the dogs raced along the river bank, dashing off wildly into the jungle in pursuit of fresh scents. Suddenly there was a mad yapping from Kerebau, the lead dog, and a large wild pig rushed towards the river. Seeing our boats he turned but was immediately surrounded by barking dogs, who rushed in, snapping and biting at his legs. Sipi charged at the cornered animal, spearing it before it could turn to face him. I was greatly impressed by the teamwork of man and dogs but regretted our good fortune the moment the pig's considerable bulk was added to our already immobile vessel.

It was after six when we finally reached our destination at the mouth of a broad stream, Lungtudtul, and set up camp in the fading daylight. The tributary twisted and wound its way through a steep and treacherous valley. A few miles upstream from our camp Lungtudtul was blocked by a great landslide and a broad lake had formed behind the barrier. Dead trees floated on the still water and the silent atmosphere of death hung over the whole valley. For over a mile the landscape was scarred by a wide, bare, treeless avenue of white rock rubble.

Although there were nests and traces of orangs having fed these were a couple of weeks old. I met only three orang-utans in ten days and they were merely passing through on their way to the west. Nor did I have any more luck in the adjoining valley of Pederesan. A male was calling but I was unable to find him. A small tree richly covered in wild plums lay in my path. The juicy red fruit were sweet and delicious. I pushed on through the thick undergrowth resolving that I would collect some of the fruit on my way home. Suddenly I was overtaken by a sharp rainstorm and the sounds of the forest were soon blotted out by the blinding curtains of rain. I crouched beneath a leaning tree to wait out the torrential downpour but was soon soaked to the skin. Cold and miserable, I abandoned my search and retraced my steps. As I reached the plum tree I saw to my horror that it had been stripped completely bare. I had been away only two hours but every fruit was eaten. Yet there was no sign of the rain-drenched orang-utan who must still be close by.

Disheartened, I plodded wearily homewards to find that Sipi and his dogs had not been idle in my absence. They had killed another two pigs and we feasted for a week on the charred fatty meat. I became somewhat bored with this diet but there were no catfish in the Bole and the men refused to fish for the less tasty varieties. After one unfortunate fish stew and endless mouthfuls of tiny, needle-sharp bones I had to agree with them.

We stayed up Bole for only two weeks before I became discouraged and determined to return to find what the Segama could now offer. With Tulong's boat finished and able to transport half the luggage the trip down the swollen river took less than three hours. Back at our base camp I found a message from the returning Iban geological team. The *Javi* script was inscribed in charcoal on a large block of wood. Sipi was able to decipher the Arabic characters and told me they advised me to hurry back to the Segama, where great luck awaited me. Another note from Meng told me that during their trip they had seen several orang-utans. As a parting shot he advised me to eat monkey meat if I suffered from any more fevers; it was claimed to be a powerful remedy.

The Ibans' prediction proved correct for, although the area to the north of the Segama was still deserted, I found many recent traces of orang-utans on the other side of the river. Here the ground was moist and the forest still fresh and green. Twice I found animals there and decided to tackle the region more seriously. Since it took several hours to reach this point from our camp I could do little useful observation in a single day. I decided, therefore, on a six-day visit and set out loaded with food, a tape-recorder and my precious polythene sheets.

My trip was a huge success. I found a family of three orangs and was able to watch them almost continually for two days. They led me far from my base and at night I heard calls even farther to the south. Travelling towards them I encountered several more orang-utans and was soon making up for the lean period of the previous month. I returned for another long trip, again sleeping out with my subjects so that I could keep them under constant surveillance. Every orang seemed to be on the move and the entire population continued to migrate south to a moist plateau, the source of not only Little Bole but Lungtudtul and Pederesan as well. Here I bumped into several animals each day and occasionally as many as five together.

Fruit seasons were very local with the various trees cropping at different times of year and the orangs moved around following their favourite foods. While fruit had been abundant in the lowlands the orangs had been scattered throughout the area but now that the weather was drier they had been forced to fall back into the hills where fruit and fresh leaves were still available. Because of their great bulk and arboreal habits orang-utans can forage only a small area each day. Consequently they prefer to travel singly or in small groups. Even when orangs were crowded together, as now, and different family groups met frequently, they still maintained their haughty distance and did not show any of the friendly greetings that characterise more social apes and monkeys.

Likad returned to camp but as I had developed a liking for pig-meat I decided to keep on Sipi and his dogs as well. In fact this proved to be rather a mistake because both Sipi and Likad now decided to make boats like Tulong's, and since this work took up all their time they were rarely available for other chores. The work was fascinating to watch, however. First they chopped down a large, red *Seriah* tree and cut off two lengths for their boats. These were crudely shaped with an axe before the inside was carefully hollowed out by adze. The geometry involved in this was very complex, especially as all measurements were made with short lengths of twig. Sipi was the master boatmaker and, under his supervision, two very fine *sampans* began to emerge. Both Sipi and Likad were very rude about poor Tulong's effort, which was still very unstable in the water, and called it 'the floating coffin'. When the boats were nearly finished they had to be dragged several hundred yards to the river. By some grave miscalculation I happened to be in camp on the appointed day and found myself heavily involved in hacking a path through the undergrowth and making smooth rollers for the boats to pass over. At last we were ready, and, after a couple of hours of sweating and pulling, we had them afloat and towed them back to camp.

The great labour was by no means over. Likad's boat leaned slightly to one side and had to be worked further to make the wooden sides of equal thickness. Then the boats had to be stretched open. They were first hammered all over, inside and out, to prevent the timber from cracking; then they were raised above a long fire and the underside was roasted while the men poured water to keep the interior moist. This took several hours and had to be done in the humidity of the night. Slowly the boats curled open like ripening pods, becoming broader and shallower until their owners were satisfied they would be stable in the water. Only if the boat is cut to the right shape will it open wide and retain a level keel. Bamboo seats were fitted into place and bound tightly with *rotan* to prevent the sides from curling back again. At last the two craft were in the river and ready for use but now we had too many boats for us to man.

Although busy with their boat-making the men had not ignored the other offerings of the jungle. Each time I returned to camp after a few days' absence I would find more wares accumulated. They had collected several miles of *rotan* cane, much of this already woven into baskets and cage traps for fish. We also had an enormous pile of carved *belian*-wood boat paddles and several large bamboo containers filled with pig fat and preserved meat. There was certainly no question of our ever being able to move camp again.

Time was running short and I would soon have to return to civilisation, but before the trip ended I hoped to obtain some better photographs, or even film, of the orangs. As I had not wanted to carry heavy cameras with me in the forest or expose them too much to the humid atmosphere, I had limited myself to photography only when animals were near to camp. I decided to make a last effort and amassed enough supplies to last me for ten days. With my cameras and tape-recorder I had more than I could carry, so Likad and Tulong helped me transport the equipment into the jungle. They were amazed at the distance we travelled and had never imagined that I came so far into the forest. As they went they bent over the tops of saplings to mark their way back. It took us about four hours to reach the spot where I had seen orang-utans a week earlier. The men made a raised sleeping platform and fitted the polythene sheets overhead. They hurried off home, terrified that night would sneak up on them before they reached the river. I made a brief survey of my surroundings but, alas, the orangs seemed to have moved on again.

I searched to the south as far as Lungtudtul and to the west into the high mountain ranges but found no sign of my red apes. Only after three days of weary footslogging and twice moving all my food and equipment to new campsites did I eventually locate a small family returning northwards. Molly was a large, dark female accompanied by a small, orange juvenile, Monty, and her little chocolate baby, Mitz. At midday Molly built a nest for herself and the infant. They played for a while before lying down to rest. Monty left the sleepy family and moved off determinedly. He travelled quite fast but the undergrowth was thin and

I could easily keep up with him. He arrived at a large tree and climbed down to a hole in the trunk. Hanging by his feet and one arm, he peered into the hollow, inserted his right arm and started to scoop out water. Every few minutes he paused to look around, gazing surreptitiously to right and left. He scooped the liquid rapidly, sucking the droplets from the hair on the back of his hand. Water splashed over his face and dribbled down his chest. Twenty minutes passed while he quenched his thirst, then, satisfied, he climbed up to rest in the branches above.

This behaviour was typical of Monty, who often asserted his independence by deserting his mother and small sister and wandering off alone. He always rejoined them in the evening and would then play late with Mitz long after Molly had made her nest for the night. Eventually, when it was dark, a tired little Mitz would creep into Molly's warm bed while Monty made his own small nest a few feet above them.

I stayed with the family for six days, following them as they moved slowly down one of the branches of Little Bole back towards the Segama. At night I could hear elephants in the distance but none came close. Often as I lay on my uncomfortable bedding I was visited by a large spotted civet, who came to lick clean the tins from which I had eaten. I would hear him rattling about and shine my torch to find two round eyes glowing red among the shadows. He never showed the slightest fear of my presence.

One morning loud crashings heralded the arrival of a long-faced old male, Nicholas. He was a splendid character and I was able to take some nice cine film of him feeding, nesting and shaking branches at me in vigorous display. Nick proved to be only a passing visitor but Molly and her offspring remained in the vicinity, diligently munching their way through a tree full of unripe *bubok* fruits. Not even a heavy shower of rain could deter them from their task and I sat watching, cold and damp, beneath the dripping leaves. I heard noises of animals approaching on the ground behind me but, presuming them to be pigs, I paid no attention. When they were only yards away I turned to find, to my horror, that the new arrivals were not pigs but orang-utans. A sub-adult male led the way, followed by a juvenile and a female. I stood up grunting; they stopped, as surprised as I had been, then climbed hurriedly into the trees to watch me. It seemed I had hit the centre of a party, for within a few minutes another two orangs appeared. The three families paid no attention to one another but fed in neighbouring trees till dusk, making up for time lost during the rain. A large moon came out and I could still hear them moving and feeding, well after I had crawled into my polythene chrysalis for the night.

On the last morning of my trip Monty again left Molly and set out on one of his lone excursions. Half an hour later he met another family and hurried towards them. A juvenile of his own size joined him in a terrific game of swinging, biting, wrestling and chasing. I thought I recognised her but was not sure until I saw her mother resting in the tree behind. They were a couple I had seen a fortnight earlier nearly two miles farther south. It was a marvellous conclusion to my study and I was able to take some good cine footage of the two youngsters swinging gaily together in the lianas. Then, bidding the orangs farewell, I packed up my belongings, much lighter now without any rations, and started off for the rendezvous with my men.

With all the jungleware we had managed to collect, our return to the *kampong* proved to be a major undertaking. Eventually, and to everyone's satisfaction, we solved the boat problem by tying three together to make a raft. While Tulong and I steered this cumbersome craft, carrying most of the equipment and a cage full of live *tarpa* fish, Sipi and Likad took the other boat and the dogs. We drifted peacefully down the Segama, heading at last for our respective homes. Although I had lost a lot of weight and badly needed a long rest, I was happy with the way things had gone. I had found unbelievable numbers of orang-utans and achieved over two hundred hours of unique observations.

Unlike the other apes and monkeys that had been studied, I had found the orang-utan to be a solitary nomad. His large bulk and relative immobility forced him into a hermit-like existence for few trees could bear the ravages of more than one red ape at a time. By travelling alone or in small families the orangs were able to find and make good use of the widely-scattered food trees. How they knew which parts of the forest would be productive during the various seasons remained a mystery. I suspected that the other animals took advantage of the old males' greater experience and located feeding areas from the patriarchs' calling. Alternatively, these loud broadcasts might be solely to warn other males to keep their distance, thus helping to space out the population. There were certainly many questions still to be answered and I was determined to return as soon as possible to tackle them.

Part Two

Borneo 1969-70

Chapter four

Half a Mawas

Back at Oxford I had to finish my zoology degree, but as soon as my final examinations were over I could start planning for another trip. I had been fortunate enough to receive a Leverhulme Scholarship from the Royal Society and this would enable me to spend another year studying orang-utans in the Ulu Segama.

The problems of maintaining myself for a year in the jungle would be rather different from just making do for three months. I was determined that camp should be more of a home and less of a deprivation. This was particularly necessary as I would arrive shortly before the winter monsoon when the rain and cold would make work in the forest both difficult and unpleasant. Camp would need to be sturdy and stormproof and I promised myself the additional luxuries of a camp bed, mosquito net and radio to let me know what was happening in the outside world. I also insisted on funds for an outboard motor which would considerably increase our mobility, especially when the rivers became swollen with winter rain.

Exactly one year to the day since I had left the Segama after my first trip I was heading back up-river again. The new *sampan* pushed eagerly forward, powered by the small motor, and as Likad and Tulong had been unavailable I also had new companions; Pingas, a young bachelor, and my boatman, Bahat, an experienced river-man and father of seven. My stay in England flitted to the back of my mind like a half-forgotten dream, or another life, and I felt as though I had never been away.

The gravel beach at the mouth of the Bole River was just as I remembered it but the bushes and grass had grown tall over my original campsite. We spent the first few nights sleeping on the beach until the men could clear a permanent site. Since I intended to stay for a long spell we needed to be well above the flood level of the river. Bahat and Pingas set about cutting trees at the forest's edge to make a small clearing for our three dwelling houses. Bahat wanted to bring his wife and family to live with us and this seemed an excellent idea as it would give our home life greater normality. We would, in effect, have our own tiny *kampong* far up in the Ulu. Bahat's wife would wash clothes, cook and look after the camp whilst the men were away fishing or hunting and I was off in the forest with my orangs. The new canvases I had bought did not prove as rainworthy as we had hoped, and as the monsoon season was approaching and I had a good deal of valuable and delicate equipment to protect, I sent Bahat back to town to buy sheets of corrugated iron. A few days later he returned with Sherewan, his wife, and five of their children. He had brought his own boat along, too, and both were laden with metal sheets.

The eldest son, Dainal, was blind but was a strong oarsman and proved extremely adept with his hands. He worked willingly, collecting wood, making fires, washing, weaving baskets and carving wooden handles for knives. His younger brothers, Shingit and Latig, helped Bahat fit the roofing into place and dashed about proffering nails at the required moment. The two youngsters, barely more than toddlers, helped their mother in the kitchen or spent their days splashing and playing in the river.

Rather as I had expected, there were few orang-utans about for the first month. I found only three on the north side of the Segama and an equal number on the other bank. Apart from the excitement of finding orangs again after so long away, the greatest thrill of the renewed study was the realisation that the second group I met were none other than my old friends Margaret and Midge. Midge was bigger now, but still the same button-eyed clown I remembered, and Margaret's piercing, bright eyes would have given her away even if she had not greeted my appearance with a loud series of 'lork' vocalisations, as only she could make them.

Surprisingly the fruit season was very different from the previous year. Many trees that had fruited then had not cropped again and other trees that had been barren now littered the floor with their shells. Not only individual trees had suddenly become productive, there were whole new species that I did not recognise. Amazingly there was no sign of mangosteens, *rambutans*, langsats and *mata kuching*, fruits that had been the orang's favourites the year before. But if the forest was different in some ways it was surprisingly unchanged in others. The trails I had made were still quite clear. Nothing had grown and where I had lopped saplings, the tops hung down and the leaves were still green. The same dead logs lay among the leaf-litter. Of our own camp there had been no trace but I could make out the bent branches and black, leafy remains of some of the orang nests I had seen built a year before. Most amazing of all was a broad, dead *belian* trunk bearing the scars where someone had cut out large slices of wood for boat paddles. The wood chips lay on the forest floor exactly as I remembered them. Incredulously I told Bahat about my find and he explained that the flakes had lain there not for just one but over twenty years. Bahat himself had been one of a party of Dusuns who had fled from the Japanese during the Occupation and made their home here in the jungle. This was the *belian* tree from which they had cut their paddles. The wood flakes were too heavy to be washed away by the tropical rain and too hard to rot in the humid warmth. The Dusuns had certainly chosen excellent wood to get good use from their paddles.

As though at a pre-arranged signal the orang-utans suddenly returned to the lowlands. For several days there had been no sign of any, but at the beginning of November several small groups moved into my area from the west. The leader of the first band was a splendid, large male I named King Louis. I followed Louis for four days and we both spent most of our time sheltering from the terrible rain, Louis under piles of branches and myself under a canvas cape. Whenever the rain eased Louis climbed out to feed on acorns, making such a loud crunching noise as he bit them open that I could hear him from well over a hundred yards. He spent a lot of time in the area Harold had occupied the previous year and even used the same routes, but his swollen face flanges and longer beard convinced me that he was a different animal.

Louis did not like me and made no attempt to hide the fact. On the first day he was rather shy but by the second he came crashing towards me, descending through the branches as he drew close. I backed away nervously but this gave his courage such a boost that he was soon down on the ground in fast pursuit. Subsequent encounters became rather a trial for, now that he knew I was afraid, he charged whenever he saw me, and, unwilling to call his bluff, I inevitably made a hasty retreat. For observation purposes this was hopeless and something had to be done. I chose a moment when Louis was resting low down on a convenient bough. My large telephoto lens seemed the most frightening object to hand so brandishing this like a gun, and trying to look as fierce and determined as possible, I crashed noisily and directly towards Louis. As I reached the bottom of his tree I pointed the lens straight at him. Startled by my new-found courage, Louis climbed slowly upwards, peering suspiciously at my enormous eye. Pressing home my advantage, I started to climb after him and, seriously worried now, Louis clambered on to the topmost bough. Satisfied, I jumped to the ground, collected my lens and went to sit a few yards away.

For the next hour Louis remained sulking at the top of the tree like a naughty schoolboy who has been severely scolded. Meanwhile I casually looked about me, pretending I hadn't the slightest interest in anything so insignificant as a timid orang. Somehow we came to a mutual agreement that he would stop coming to the ground to frighten me if I would stop climbing into trees to harass him. When Louis moved off for his evening meal he paid no further attention to me although I remained close behind and afterwards he always seemed perfectly at ease in my presence. Since we had reached such an amicable relationship it was unfortunate that he proved to be only a visiting nomad for when he moved off quickly the following day it was the last I ever saw of King Louis.

The fruit on the *bubok* tree behind my house began to ripen and it was this sweet bait that attracted Desmond, the only orang sufficiently intrepid to venture into our camp. Desmond was a fine young male with long, chestnut hair and he broached the tree late one evening. It was long after dark when he finished his dinner and settled down for the night in a large nest. This seemed an ideal opportunity for some photography, but before first light Desmond slipped away from camp and returned whence he had come. His next foray on to our territory resulted in a spectacular confrontation. Returning to camp I heard a terrific uproar of barking dogs and squealing orang. I rushed up to find Desmond trapped in an isolated tree, surrounded by three yapping dogs. Angrily he climbed down towards the excited animals and I hate to think what would have happened if I had left them to battle it out. Not wishing to witness a fracas, I pelted the dogs with bits of wood and as they scattered Desmond seized the opportunity and hurried off across the clearing to the safety of the forest. It says a lot for *bubok* fruit that, despite his experience, Desmond was back in camp feeding happily in the fruit tree a few hours later. This time he demolished all the remaining titbits, thus removing the need for another dangerous visit.

Like Desmond other large males frequently had to come to the ground for travel, but the females and young orangs were able to live virtually independently of the forest floor. These lighter animals did venture groundwards occasionally, however, to obtain small quantities of mineral rich soil. At the north end of my research area stood a limestone block sheltering a pile of hard, reddish earth. Imprints in this soil were unmistakably the toothmarks of orang-utans. Every month or so a new set would appear, evidence that another orang had visited the lick. Analysis of the soil showed it to be rich in sodium and potassium, which are rare elements in the acid forest humus and presumably of great value to the large mammals. The fact that orangs regularly visited such a local site indicated they had an excellent knowledge of the area's geography.

By the beginning of December I knew of several orangs now resident within my research area. Although the study was going well, fruit had become scarce and the weather was getting worse. There are few things more frightening than a thunderstorm in the forest. Violent winds toss the trees back and forth without mercy. Branches tear free and crash to the ground with terrifying frequency. Sometimes whole trees topple noisily to the floor, dragging smaller companions with them in a suicidal plunge, ripping an untidy hole in the forest canopy and completely blocking the ground passage. I always regarded falling timber as the most serious danger in the forest and several times had narrow escapes from falling missiles. Once while we were on the river a tall tree simply collapsed off the bank into the water ahead of us, swamping the boat before we had time to change direction.

The male orang-utans shared my dislike of falling timber and showed their displeasure with loud roars whenever they were disturbed in this fashion. When all other means of finding orang-utans failed I would simply sit in the forest waiting for a treefall and the subsequent protest to tell me whether I was in business or not.

It is the big old males who go in for such vocalisations. These high-ranking animals are up to twice the size of an adult female and their long hair, beards, high crowns, enormous inflatable throat pouches and enlarged fatty cheek flanges all attest to their status. Wherever they wander they announce their presence vociferously, warning other males to keep clear. Rivals may challenge the caller by bellowing in reply and at times the forest peace is shattered by up to three contestants shouting long-range threats at one another.

On the north side of the Segama River lived one particularly rowdy trio, Harold, Redbeard and Raymond. I met Harold up near the Tooth-Rock and he looked just as he had done the previous year, his fingers still stiff and his back still bare. I saw quite a lot of him in the next few months and as he was not afraid of me he was always a pleasure to watch. Redbeard was a more difficult customer and my first encounters with him were rather frightening. He was an incorrigible ground walker and as soon as he saw me would rush straight at me. Three times he chased me and three times I fled before I was able to repeat my tactics against King Louis and managed to bluff Redbeard into showing more respect. Between them these two split the region. Harold ranged from the centre ridge to the west while Redbeard's domain lay to the east. The single narrow ridge which served as the boundary was the scene of tremendous displays, Harold directing a barrage of calls east and Redbeard yelling westwards but they both took good care never to be there at the same time.

Raymond was a much quieter orang though very large. He did not visit the area often and showed no respect whatsoever for the Harold-Redbeard boundary ranging happily on either side and showing no fear of either of these heroes. This was rather amusing since he was quite terrified of me and would hide motionless at the top of a tall tree whenever I approached. One week when Harold was in residence on Central Ridge and had been calling several times a day, we heard the unmistakable tones of Raymond nearby. His solitary bellow had a dramatic effect for it was ten days before Harold dared give tongue again. Redbeard was no braver. When he heard Raymond calling in the distance he chorused a reply and hurried to investigate this rude intrusion. As Redbeard drew closer, Raymond called again. This time Redbeard must have recognised who it was for he turned tail and fled back the way he had come.

Obviously strong jealousies and competition existed between the males. As a spacing mechanism their calling behaviour was very effective for adult males practically never met each other. On the rare occasions when they did the encounter was always accompanied by terrific branch-shaking displays until one of the antagonists turned and fled. Probably the great size, long hair, enlarged faces and ugly expressions of male orang-utans are designed to strike fear into the hearts of their opponents. However, Nature does seem to have over-excelled herself as the female orang-utans are also frightened by their fearsome brothers and it is the younger sub-adult males who have more success with the ladies. Adult females sometimes fed near to Redbeard but Raymond and Harold were always alone. Nor did Harold desire such frivolous company, for on one occasion when he heard a female and juvenile moving towards him he climbed smartly to the ground and slipped silently off before they saw him.

I came to know some of the orangs, like Harold, quite well and learned to recognise their particular habits and idiosyncrasies. But many of the animals were temporary visitors whom I would see for only two or three days before they left my part of the woods for good. Meanwhile I had established my own range with my own system of paths and my own favoured haunts. There were places where I habitually drank from streams or cut refreshing vines for their sweet sap. I had special lookout points and listening spots, favourite log benches and tree blinds. With the help of Bahat and Pingas I erected several polythene-covered shelters at convenient sites throughout the region. Here I stored caches of tins so that food and shelter were always at hand wherever I might be working.

Every part of the forest was filled with memories of previous experiences, good or bad. Gradually I became a part of the jungle and felt at home there. I no longer hated the thorny *rotans* for I had learned the usefulness of their long, supple canes; bark and vines I could use to bind my shelters. I could turn the thickness of the undergrowth to my advantage, using it as cover to creep up unseen on the orangs so as not to disturb them. Moreover, the forest provided me with a tasty diet of fruit and edible fungi. My whole attitude towards the jungle had mellowed. It was no longer an enemy but a friend. Even the leeches had their uses. Tied to a creeper tendril and lowered into a stream pool, they proved irresistible to the small fish. By the time a fish realised this tempting morsel was not a worm and tried to spit it out again, the leech's suckers had held fast. A tasty snack of toasted fish of a quiet evening made a welcome addition to my otherwise monotonous rations.

I had become used to my own company and was no longer bored or lonesome in my solitary life. Sometimes I treated myself to a running commentary of my progress or soundly berated the leeches as I detached them from my legs. I was following rather solitary nomadic animals so had become a solitary nomad myself. I travelled when my orangs travelled, ate when they ate and slept when they slept. I was reminded of the story of the man who stayed too long in the jungle and turned into an orang-utan; I wondered if I wasn't becoming half a *Mawas* myself.

Chapter five

Pig-Rain and A Treasure Cave

Black clouds were building up to the south and the treetops were already astir in anticipation of rain. It was still early, but, having no desire to be caught in a winter thunderstorm, I headed homewards. Back at camp only a fine mist of light drizzle hung over the river and all was activity. Bahat, Pingas and Shingit were busily collecting their oars and our long, broad-bladed spears. I asked Bahat where they were going. He pointed upstream with his chin and replied, '*Hujan babi* (pig-rain), tuan.'

Hastily dropping my jungle pack at the hut, I grabbed my own black *belian*-wood paddle and rushed after them to the beach. We climbed into the smallest boat and paddled our way slowly upstream against the swollen river. Two of the dogs, keen to join us, ran along the shore barking excitedly. Eventually, discouraged by the rain and our refusal to have them aboard, they dropped back and returned to camp. Silently but steadily our small *sampan* crept along, half hidden beneath the thick vegetation of the muddy river bank. The men pointed out the flattened ferns where migrating bands of pigs had made their way from the forest to the river's edge and the deep tracks on the other side where they had scrambled ashore again after their hazardous crossing. After an hour of hard paddling we reached Pulau Pin Pin, a small shrub-covered island in the middle of the river. A hundred yards upstream of a small gravel beach, we pulled our boat under the overhanging branches of the opposite bank. As we passed we had seen signs that several families of wild pigs had already crossed from this spot but Bahat assured me that pigs would continue to migrate north for another two months.

For twenty minutes we waited quietly, then Shingit gave a silent jerk on the side of the boat and we all crouched low. At first there was nothing to be seen but I could hear the low grunting of pigs in the distance; then there was a slight movement among the tall ferns and the pigs trooped on to the beach in single file. A large boar moved forward, then turned and lunged savagely at a younger male who had come too close, challenging his authority. The latter squealed and retreated and the old leader again approached the river's edge. He raised his long-bearded head and scented the air to left and right, then peered suspiciously through the rain to the other bank where we lay hidden. The white of his long tusks glinted as he turned and, with a loud derisory snort, trotted away from the river through the bracken, the rest of his herd following at a respectful distance. Ten minutes later we saw them swimming speedily and safely across the river farther downstream. Bahat cursed the old boar's caution and we settled back to continue our vigil. A few minutes later another band of pigs crossed but again too far away for us to reach them. With amazing speed they reached the shore, climbed up the slippery bank and disappeared into the forest.

It was already beginning to get dark when I felt another jerk on the side of the boat. We crouched low, as still as dead folk, while the ten pigs tiptoed their way across the beach and lined up by the river. An old boar waded out into the stream, then plunged forward and the others followed in quick succession. We nosed our boat out from the branches and paddling furiously headed it into the fast currents that would bear us down on the swiftly swimming pigs. The boat was moving rapidly but the pigs were already halfway across the river. We paddled madly faster and faster. Bahat suddenly changed course to head the pigs away from the landing point. They turned and headed downstream, giving us the vital seconds we needed to draw close. As we swept down on them the leaders were already clambering up the bank and crashing through the undergrowth but the tail-enders were still afloat. Bahat lunged forward, spearing a pig in the river, and I threw my weapon to hit another half way up the bank. It tumbled back into the river, where Pingas, leaping from the boat, grabbed its tail. The beast tried to turn and face him but Shingit speared it clean through the heart. Bahat was already cleaning his pig and Pingas and Shingit started on the other while I brought the boat up to the bank beside them. As night fell we drifted home, well pleased with our evening's work and joking loudly, while Bahat sucked happily on a damp cigarette rolled in a palm leaf.

No-one knows quite why the pigs migrate, where they come from or where they go. They do not migrate every year nor always in the same season but, as Bahat had predicted, for the next two months there was a continuous flow of them moving north as though the Devil were after them. Many of the herds were accompanied by squealing, striped piglets, only a week or two old, and all the animals we killed were fat and in good condition. The men were kept busy hunting every day, while back in camp Sherewan and the children cut up the meat and fatty skins ready for frying in a large, round-bottomed pan. The meat and oil were packed into Chinese biscuit tins, bound with *rotan* canes, and stored under the houses. Preserved in the oil the meat remains edible for several years and by the end of the two months the men had built up a large pile. Traditionally pig fat used to fetch a high price in the towns, but the spread of the Muslim faith and the cultivation of oil palms in Sabah has greatly reduced the market and our supplies would go no farther than the Dusun villages.

During most of the year visitors were rare so far up-river but as long as the pigs continued to cross the river there were boatloads of Dusuns travelling up from the *kampongs* to lie in wait for them. The rain-swelled river enabled the large crocodiles to move downstream from their pools with similar predatory intentions. Several times we spotted their menacing shapes among the débris floating past our camp and sometimes half-eaten pig carcases drifted downstream, bearing terrible rends where the crocodiles had ripped them to feast on the rump and bowels.

At the end of December I was laid up in my hut for three days with septicaemia of the leg, but on Christmas Day I was able to walk again. As there was no fresh food left in camp we set out in the boat to look for something suitable for the festive occasion. After a few miles of travelling west we reached Butok Batu, where the river turns sharply left, and from there on we paddled southwards.

We surprised a family of otters playing on the beach, rolling and biting, chasing and sliding till one cub spotted us and gave a shrill whistle. Five black bodies snaked away up the bank and vanished into the forest. Some while later we rounded a bend just in time to see a boat of Dusuns paddling hastily away from what appeared to be a large boulder in the middle of the river. As we drew near we found to our surprise that it was a freshly-killed female elephant. A small hole in her side where the lead balls of a home-made cartridge had penetrated the thick skin and several spear gashes left us in no doubt as to how she had met her fate. Uprooted bracken and earth had slipped down the bank where she had struggled in her death throes. I was amazed at the effectiveness of the home-made armaments and the bravery of the men who, so poorly equipped, dared to tackle the most powerful animal in the forest. The elephant, however, is protected by law and there was no sign of the poachers when we rounded the next bend. They were obviously hiding up in one of the small streams in fear that I would report them to the game warden. They had not even had time to take meat from their trophy.

I spent the afternoon trying to photograph two elusive crocodiles in a deep river pool but all I saw of them were their eyes and nostrils. Towards evening we drifted back to the huge beast and helped ourselves to several generous steaks, welcome supplies for our larder. Later that night we sat around the camp-fire eating our far from traditional Christmas dinner. I was surprised at the tenderness of the flesh and Bahat assured me that there was a ready market for elephant meat in the town as no-one could tell it from venison.

Over the months I sampled many strange creatures, some pleasant, others less so. Porcupine was delicious and so was the black, tree monitor lizard that lived in the forest. The huge river monitors, however, were very tough and had to be stewed for several hours. Often we feasted on terrapins, tortoises or frogs. There was nothing wrong with the meat of a leaf monkey caught by the dogs but I was rather put off by its yellow fat and was once horrified to find that the entire head had been left in the pot. Perhaps the greatest delicacy were the eggs of the *labi-labi,* the large soft-shelled river turtle, that we dug up on the sandy beaches. I kept one clutch of turtle eggs in my house and eventually forty little hatchlings emerged. I spent several hours experimenting on how they found their way back to the river, a task at which they proved very competent. By following downhill gradients or, if in doubt, heading for the brightest area of sky they invariably met the river before too long. Sherewan became very fond of the turtles and would cradle them fondly to her bosom. We accommodated them in a large water tub and at night Sherewan would take the tub and all forty occupants to bed with her until Bahat firmly put an end to the habit. As the turtles outgrew their home we returned them to the river and their fostermother was very sad to see her pets swim down to the muddy bottom and burrow out of sight.

Sherewan was a keen gardener and we soon had enough maize, sweet potatoes, chilli peppers and cucumber to serve our needs. On top of this she also brought a fine pair of chickens up from the *kampong*. Every three hours throughout the night the cock would crow and my hatred for him grew fast. There were no other cocks to give reply but an eagle nesting on the other side of the river struck up a curious relationship with our crazy bird and responded to his crowing with discordant cries. One day our cock was nowhere to be seen and as he did not return on the following day we feared him dead. On the third morning, however, we were woken by a distant crowing on the far side of the river. It took two more days to catch the wretched bird, who had become quite wild, and when we finally retrieved him he was battered and bruised and lacking many of his glossy tail feathers. Since he couldn't fly we never knew how he had crossed the river but I always suspected his eagle friend of having a hand in the matter. In spite of these adventures the cock took his family duties seriously and his wife produced several clutches of eggs which hatched into cheeping furry bundles. No sooner had each batch of chicks emerged, however, than its numbers would be steadily depleted, white chicks first then brown, as the snakes and monitor lizards took their toll. Eventually, dispirited, Sherewan collected all the remaining chickens and took them back to the village so that we were able once more to sleep in peace at night.

Bahat never took sugar in his tea but he was passionately fond of the honey of wild bees. A honey-collecting expedition was always preceded by a week of frantic activity. As soon as he found a tree where bees' combs hung, Bahat would begin collecting *rotan*, the long climbing palm stems from which chairs and baskets can be made. Three hundred yards of *rotan* were needed for a honey trip and were plaited together to make a strong rope for the large collecting tin. In addition about two hundred bamboo pegs had to be made, each one taking several minutes to cut and shape. Other necessary items included a strong *belian* wood mallet, countless pieces of strong liana fibre, thirty cut poles with shaped ends and a finely shredded then retied length of dried liana for a torch. Often Bahat would also build a special little house at the scene of operations for the great day.

We were all ready for our first honey expedition. Pingas and blind Dainal rolled up the *rotan* rope on the beach, while Bahat and Shingit gathered the bamboo pegs, collecting tins and other equipment. We left the dogs behind and crossed the river near Little Bole. Twenty minutes' fast walking along a fresh path Bahat had cut for the occasion brought us to the foot of a huge *Mengaris* tree that towered some two hundred and thirty feet into the air. A small hut and the cut poles were already waiting and Bahat and Dainal immediately set about constructing the ladder. Pingas and Shingit cleared the undergrowth all round the base of the tree so that we could move about easily come nightfall to retrieve the honey as it was lowered. Far up above us in the tree hung the eight, large, pendulous combs which were the cause of all the activity.

Bahat built a firm platform at the base of the tree and the ladder progressed slowly. Blind Dainal felt the trunk for a suitable anchorage and wielding the *belian* mallet drove a bamboo peg into the wood. When the point was embedded only two inches, Dainal took the upright pole and bound it firmly to the peg with liana fibre. He climbed up one rung to repeat the task. It was terrifying to watch this blind man now halfway up the smooth trunk, calmly hammering in pegs and building up the ladder. As he came to the end of each pole another was passed up to be lashed to the last and again tied to the little pegs. These were fixed every three feet but alone could not carry the strain of a man standing on them. Only the combined strength of all the poles and pegs beneath give sufficient support, and the climber must hold on to the poles, rather than the pegs, if the whole thing is not to swing away from the tree and collapse. I wondered whether Dainal could conceive of the terrible void beneath him and decided that perhaps it was an advantage to be blind for such a job. When he reached about a hundred feet Dainal retired and Bahat took over the frightening work. Eventually he reached the first branches of the huge tree and he too climbed down for a rest.

It was already dark but Bahat wanted to wait for the moon to go down so that the bees would not see him. Well after midnight he returned to extend the ladder on to the branches and after another rest he began the final assault on the nests. This time he carried the end of the *rotan* rope and a burning liana wick which glowed like a giant cigarette. Near the first comb he suddenly swung the glowing torch wildly painting weird patterns in the sky. The wick burst into flame and when he knocked it against the nest it exploded in a spray of sparks which tumbled to the ground amid the buzzing of thousands of angry bees. We dashed for safety in the dark as the bees flew for the fire in a desperate suicidal attack. Shrieks and howls from Bahat indicated that despite all precautions some of the bees had found their molester. Ignoring his injuries, he cut the comb with a wooden blade and filled the four-gallon collecting tin. He lowered it safely to the ground, where we emptied the precious contents. So the night continued, the tin being hauled up and lowered again, heavy with comb and honey and grubs. Bahat attacked the other nests with more fireworks displays, getting stung again and again. I was thankful for my boots as the ground was alive with lost and singed bees and both Pingas and Shingit had been stung.

By two o'clock we were through. Bahat lowered the last tin of honey and climbed down to join us. We hurried off to the boats and were soon back in camp with our booty. Poor Bahat's face was so swollen from his stings that he could hardly see and I gave him two strong antihistamines for the night.

By breakfast time he was fully recovered and happily gorging himself on the waxy honeycomb, grubs and all. I was offered a huge bowl of this sickly cereal but could manage only a little clear honey. It was certainly delicious, quite unlike any domestic honey I had tasted, but was too sweet to eat in large amounts. Soon we were surrounded by swarms of bees who had picked up the scent and informed their friends of the good news. We had collected some four gallons of clear honey but even so it seemed an incredible amount of work, effort and danger for the reward and made me realise what value must have been placed on wild sugars before man cultivated the sugar cane. Probably the only sweetness known to Prehistoric Man in Borneo was honey and ripe fruit.

Bahat was not the only creature in the forest willing to risk his skin for wild honey. *Baruang*, the sun-bear, earned his living in this way and the tree-trunks were pitted where his sharp, curved claws had gripped as he climbed to the combs. Even the orang was not impartial to bees' nests and I once witnessed the amazing sight of an old male who had somehow managed to reach the top of a *Mengaris* tree. Sitting above the nest, he munched busily at handfuls of torn honeycomb. He was surrounded by an angry swarm of bees but, apart from waving them away with one arm and keeping his eyes tightly closed when not locating another juicy morsel, he did not seem at all worried by them.

My important role as camp doctor seemed rather ironic in view of my own frequent illnesses. Usually my duties involved prescribing aspirins, malarial tablets or worm purges but sometimes more skill was needed. On a hunting trip Bahat was badly gored by a pig and streaming blood he was carried back to camp. The pig had bitten a gaping tear in his arm and Bahat was very ill indeed. He could still flex and extend all his fingers, however, so I presumed no vital nerves or tendons had been severed. Having drugged him with tablet morphia, I set about the awesome task of sewing him together again. I bent and sterilised a sewing needle and sharpened its point. Fine nylon strands from a piece of rope served as thread. After several trials I found a knot that would slide tight easily but not work loose again. Then, extremely nervous, I put five stitches in the L-shaped gash. Bahat remained very ill for several days and I kept him high on antibiotics to prevent the wound from becoming infected. At last it healed and I thankfully removed the stitches. All that remained of the injury was a slight swelling and an area of skin dead to pain and this proved a great success at subsequent parties. Bahat congratulated me on my steady hand and described in graphic detail the tremors of the doctor in Tawau who had sewn him up when he was savaged by a bear.

For common ailments the Dusuns have their own remedies, many based on superstition but some possibly of sound chemical origin. Bahat and Sherewan were for ever collecting plants that were good medicine and Pingas was very keen to find eagle eggs that, he said, were powerful magic. Some of their methods seemed to work. Certainly the dogs always showed a marked hunting improvement when their noses had been held over the smoke of burning herbs and they had been rubbed gently with feathers.

While I was visiting the *kampong* on a trip to town a ceremony was held to cure a man of fever. Many of the village turned up for the performance, which turned into quite a social occasion. The sick man lay at one end of the house while the head magician and two disciples lit fires and burned roots, chanting unintelligibly all the while.

They raised their colourful robes to cover their heads and performed a wild blindfold dance to the rhythm of the beating gongs. This strange ritual continued for several hours while, from time to time the chief magician broke away to rub eggs over the sick man or suck all over his body to bring out the evil spirits. Although they danced non-stop till dawn the performers appeared perfectly fresh. The patient claimed to feel much better and everyone was very happy.

While there is no fee for such a service the grateful recipient usually makes some donation, often a chicken, to the magician. It is as well to do so for he has the power to bring on sickness as well as cure it. The Dusuns always tried their *kampong* medicine first. Only when this failed to bring relief would they travel to the hospital in Lahad Datu or Tawau.

Dusuns are true animists and believe in all manner of ghosts and dreadful supernatural beings. Like other Bornean peoples they have a vivid mythology of stories and fables, handed down from generation to generation by word of mouth. One night as we sat over a meal of rice and chicken, one of the men suddenly rushed outside shouting and began to bash the metal water tub with a saucepan, making a tremendous clanging sound. Immediately the whole village took up the call and the banging spread all along the river. I asked what all the noise was about and my host explained that the *Labi-labi*, the large soft-shelled river turtle, was eating the moon and must be frightened away. I went outside and sure enough there was a large chunk missing from what had been a full moon. The missing portion grew larger and larger as the eclipse continued. At the first sign of the obscured area reappearing a loud, shrill cry pierced the night '*Buah kadil*' (the name of a type of fruit of which the *Labi-labi* is afraid) and the whole village started shouting and singing and beating their tuned gongs. The celebrations at the return of the moon and the defeat of the turtle went on late into the night and it would obviously have been wrong for me to disillusion them from their beliefs.

Dusun villages differ markedly from the better-known long-house system in Sarawak. Each family has its own small home, perhaps a hundred yards from the nearest neighbour. The *kampong* straggles along the river banks with the stilted houses set back at a safe distance in case of flooding. Even so, every few years unusual floods wash away many homes and crops and the Dusuns must start again. Around each house the forest has been cleared to grow a rich variety of vegetables; tapioca, sweet potatoes, maize, pineapples, papayas, limes, fiery chilli peppers, guavas, betel vines and betel nuts. Groves of scarlet fruiting *rambutans* mingle with fragrant mangosteens, bittersweet *langsats*, yellow and red bananas and strong-smelling durians. Chickens scratch in the dirt, and dogs, goats and pigs wander the *ladangs* in search of food. Farther away are the *padi* fields, shared by several families. Hill rice is grown as it needs no standing water and less attention than the *sawah* rice of the lowlands.

Each household has at least one boat vital for communication and for catching fish, the main source of protein. The *kampong* has a total of four *orang-tuah* or elders and one village headman or *oti*. These are the men who officiate at weddings and funerals and decide disputes between villagers. Women have many rights and retain their own property throughout marriage. Divorce is easy and acceptable. A woman seeking divorce must convince the *oti* that her husband has committed one of a multitude of village sins. If he is judged guilty the divorce is granted and the husband loses his *wang-kaseh*, or love money, which is paid to the bride's father before a wedding. However, if it is the man who seeks a separation and he can prove that his wife has behaved wrongly he can get back most of his *wang-kaseh*. The amount paid for a bride decreases according to her age and previous marriages. Since divorce is common among

Bahat, the ardent honey collector

Opposite Flying squirrels can glide over a hundred yards Below, Draco parachutes down to a gentle landing

Left Head of a poached bull banteng
Below We feasted on the meat of the wild pig

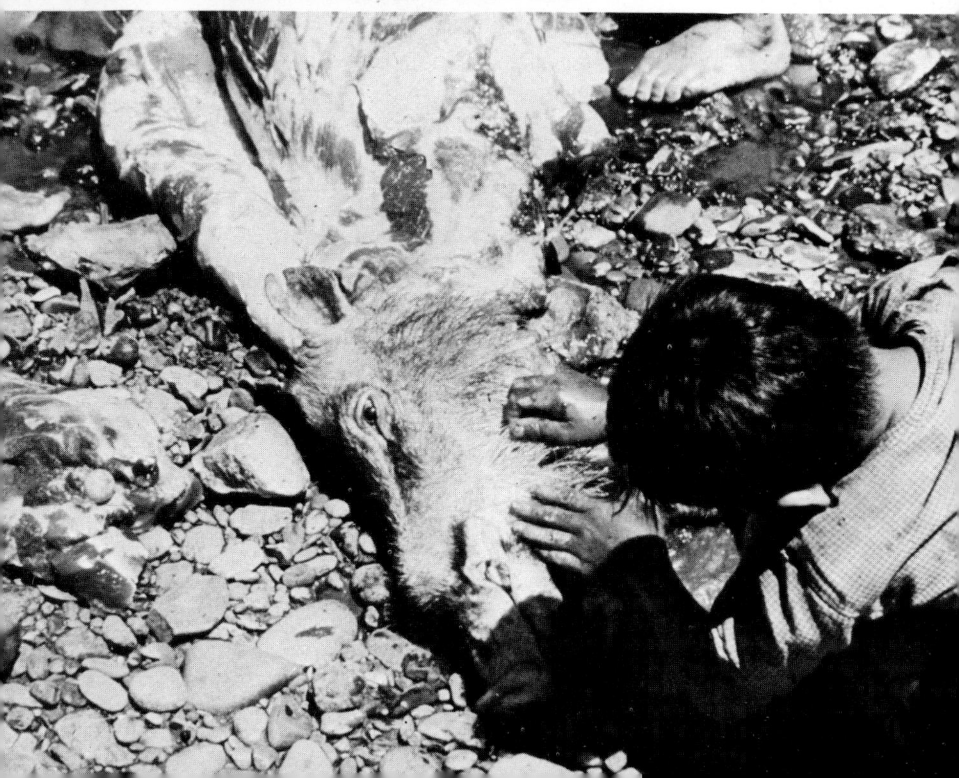

Above Seamus looked very sweet kissing my bride to be, and below, old friends get together at Sepilok

the Dusuns family relationships are very complex and all children remain with the mother even when she remarries. Bahat and Sherewan had both had previous partners and since marrying each other had been divorced and rejoined no less than four times!

One night I attended a wedding in the village. Dressed in all their finery, the bride and groom sat in a small room surrounded by the bride's proud relations. Villagers came to pay their respects before joining in the feasting and merrymaking. The women sang and chatted together while the men squatted round the huge jars of *tapi*, joking and drinking. *Tapi* is prepared from fermented tapioca roots covered by a layer of banana leaves and the jar is finally topped up with water. At first the beverage is very potent but as the evening progresses it becomes gradually more dilute as the jar is retopped. The drinker decides beforehand how much he will take and fills a tin with that amount of water. As he sucks up the *tapi* through a bamboo drinking straw a friend tops the jar up with the measure of water. As each man can keep a check of how much he has drunk there are keen competitions to see who can imbibe the most and always considerable cheating with the water.

For several months Pingas had been very keen on one of the girls in the *kampong* and, after another two months of negotiations with her father, his marriage to Sipoiok was finally agreed and a date fixed. Sipoiok herself had shown no great keenness for the match but her father was much impressed with Pingas's regular wages, an important consideration in a Dusun marriage. Early on the eve of the wedding Pingas borrowed the large boat with the outboard motor and set off down river to the *kampong*. I told him to take a holiday and bade him my felicitations for the occasion. He was welcome to bring his bride with him when he returned.

Four days passed and Pingas came back, but alone and showing no signs of the joys of matrimony. Yes, the wedding had been fine and everyone in the village had joined in the celebrations. Yes, Sipoiok had looked lovely and been cheerful throughout the proceedings. Her father had been very pleased with his son-in-law and had given the couple a small plot of land for a house. The newly-weds, however, had been on public display day and night and had not had a moment of privacy or been able to get any sleep. Pingas looked exhausted and I decided he had better take another trip to the *kampong* and get off alone with his bride. When he returned for the second time from the village he looked, if possible, even more distraught than before. I asked him what was the trouble and he told me that he could not consummate his marriage. Pingas was convinced that Sipoiok had put a spell on him because she had never wanted to marry him and from the day of his wedding he had been completely impotent. He assured me he had never had such a problem before when enjoying the pleasures offered by the Chinese girls in town.

For the third time Pingas set off downstream; this time with a letter from me to the Indian doctor in the local hospital. The doctor's reply that he could find nothing physically wrong confirmed that the trouble was psychological. I, therefore, prepared a little magic of my own. I searched through my medical kit until I found a bottle of pills for urinary tract infection. The label bore this message: 'Patients should be warned that their urine will be bright red whilst using this drug. This is perfectly normal.'

'Pingas,' I called, 'take two of these pills three times a day for the next two days. They will cure your impotence and drive the bad magic into your urine.'

The following day Pingas informed me that he thought the pills were working and throughout the next week he became increasingly cheerful. I sent him down-river again to buy supplies and collect my mail and when he returned I asked him whether his marriage was now consummated.

'Yes, tuan, a little bit,' he replied.

Pingas never did bring his wife to our jungle encampment. She stayed in the *kampong* helping her father to build the new house and planting maize and vegetables. Pingas worked happily but was always keen for another trip to town.

It was several months, therefore, before I saw Sipoiok again. But one day when I was staying in the village she came to greet me.

'Thank you for everything, tuan,' she said with a coy smile and slipped away again to her kitchen.

One morning a boat arrived at camp with the sad news that Zinah, Bahat's cousin, had died. Bahat would have to return to the *kampong* and told me that the dead woman's husband had invited me to join in the funeral rites. Interested to witness a Dusun funeral, I prepared to leave with Bahat and Sherewan next day.

Bahat had promised to kill a pig for the feast so Sherewan sat in the bows brandishing a dangerous-looking, home-made shotgun. We drifted slowly down-river, keeping the boat in the main currents to save the motor. Twice we stopped in likely places while Bahat hurried off into the forest in search of pigs but both times he returned empty-handed. At last we saw a large boar rooting among the reeds. Bahat guided the boat silently towards the shore and inched closer and closer to the busy animal. Kneeling in the prow, Sherewan lifted the gun. When we were only twenty yards away the pig raised his head and turned to run. There was a loud report and the boat rocked violently as Sherewan was thrown back but the pig dropped and we were able to continue our journey with the promised gift. An hour before dusk we arrived at the *kampong*.

There were already many boats drawn up at the muddy landing near the dead woman's house. Her four sons greeted us and gladly carried off our pig to prepare it for the feast. We made our way through the crimson flowered hibiscus bushes to a large stilted house. Serving as a step-ladder into the house was an ancient bronze cannon, testimony to the days when the villagers had to defend themselves against the Sulu pirate boats that came up from the sea. We climbed up and entered a spacious dark room, where many mourners squatted around the hardwood coffin. The coffin had been carved by adze from a single piece of *Merbau* wood and was stained from the years it had been buried under the house in readiness for the day when Death should visit the family. At first the coffin had been open for friends and relatives to see the deceased for the last time but by now, the third day, the lid had been sealed in place with melted *damar* resin.

Seated cross-legged behind the coffin, his head covered with a white shroud, was Sunal, the bereaved husband. Bahat approached and offered his condolences then Sunal turned his tear-streaked face towards me, shook me warmly by the hand and told me how honoured he and his sons were that I should join them at this time. I returned his greeting and thanked him for inviting me, then acknowledged the other menfolk from the village who were seated around us. Behind them squatted groups of women and children.

Bahat and I left the house to bathe in the river. A fire blazed beneath a huge black cauldron where our pig was being cooked. Two men staggered to unload eight-gallon jars of *tapi* from a heavily laden boat. More and more villagers were arriving so that a small flotilla of boats were now moored at the landing. By the time we returned to the house, the wick lamps were lit and a melancholy, repetitive tune was being rapped out on a set of gongs. Two jars of *tapi* were open and the night's festivities had begun.

I headed towards a corner where I recognised a familiar face and greeted my old friend, Tulong. Tulong's eyes were bloodshot and his hand trembled. He had become very fat and I could hardly believe that this was the tough, muscular character who had worked for me the previous year. I accused him of drinking too much.

'Yes, tuan, I drink a lot now, but is that so bad?'

'Perhaps not,' I replied, 'but if you want a couple of months' work to get rid of that paunch, you are always welcome to join us again.'

'Thank you, tuan, I will keep that in mind. Meanwhile, however, you must come and drink with me.'

Tulong thrust a hollow bamboo tube into my hand and ushered me towards one of the jars of *tapi*.

Mats were laid out on the floor and the women brought large bowls of rice, stewed pig and a great delicacy, *babi jeruk*, made from pieces of pig preserved in tapioca slime, salt and sour fruit and tasting like anchovies. Tea, bunches of fruit and handbowls of water rounded off the meal. Quickly the mats were cleared and reset for another session as there were well over two hundred people to be fed.

Those who had over-imbibed became quite merry and the musical gongs clanged cheerily in the background. Were it not for the large coffin in the middle of the floor I would have been hard put to tell that I was not at another wedding. Certainly no one was letting the sad cause of the occasion spoil the fun of the night. More *tapi* jars were broached when the first became too dilute. I was not used to the powerful drink and was soon suffering from a raging headache, so excused myself from further challenges as politely as I could.

The women sang monotonous Dusun songs or chewed betel nut with black-stained teeth. Hanging in *sarongs* from two long rafters were the babies, about forty in all. Whenever a child started to cry the mother would go and rock it back to sleep then set all the others swaying again as she walked past.

A large platform had been added to one side of the house to accommodate the many guests. The roof was scant and I was glad there was no sign of clouds. Rice was poured into a large wooden pestle placed in the centre of the room. Four men, each bearing a long wooden stake, stood round the block. One started to pound the grain, then another until all four were striking in rhythm, one, two, three, four, one, two, three, four. Every few minutes a girl carefully scooped out the fine flour and poured in more rice. The flour was carried off to make a type of porridge called *bubor*, which was served to the guests. The feasting and drinking continued throughout the night but I stole away early to a corner. Soon there were so many people packed on to the sleeping platform that they were almost lying on top of each other. With the tight crush and unending noise I had a very restless night.

Dawn broke misty to the crowing of a dozen cocks, and I wandered down to the river for a bathe. Breakfast consisted of weak tea or incredibly strong coffee served with the remains of the now cold *bubor*. The coffin was wrapped in colourful cloth then bound with split lengths of *rotan* cane. Since Bahat was the local expert at *rotan* weaving he was soon heavily involved in the intricate knots of the traditional decorations. Coffins for women must be bound differently from those of men but there was a good deal of argument about the exact way it should be done and it was over two hours before this important task was finished. A parasol was erected over the coffin and the tuned gongs took up a steady rhythm once more.

The womenfolk began a violent wailing and sobbing and soon the men joined in, too. The pace of the gongs increased and the mourning became more frenzied. The mood was so infectious that I, too, found myself weeping for a woman I had never met. Two men fainted and a woman became wildly hysterical with grief before collapsing in a groaning heap. None too soon a flap was opened in the wall of the house and the dead woman's sons raised the coffin. It swayed drunkenly as they lurched towards the opening but without mishap they transferred it through the hole and carried it off to the river. Other mourners followed bearing the parasol and coloured flags aloft. The gongs beat madly, shotguns were fired, more people fainted and wails of 'Zinah, Zinah' filled the air. Still shrouded in white, the sad figure of Sunal followed behind the coffin.

Two boats were ready, the first, with the heavy coffin aboard, only inches clear of the muddy water. Sunal sat behind the coffin holding the parasol over his wife's remains. His sons took their paddles and started off up-river. The second boat was crammed with relatives, one still beating a small gong. Only men can accompany the dead to their resting caves in the forest, so the women lined the beach still wailing their farewells to the beloved Zinah. Several friends joined Bahat and me on our boat at the rear of the small convoy. Then we headed on towards the great limestone caves which had served as the Dusuns' traditional burial grounds for centuries.

After an hour's journey the boats were pulled into the flag-bedecked landing of Batu Balus, only a short distance from the larger caves of Tapadong. To the beating of the gong, the coffin was raised and carried off through a gap in the cliff along a precarious twisting path. This track led into a deep clearing ringed with a high wall of limestone cliffs and roofed with overhanging branches. Sloping up some sixty feet to the mouth of a large cavern was an enormous ladder constructed from small trees and branches bound together like the rigging of some old galleon. The coffin was raised step by step, hauled by *rotan* ropes and supported beneath by many willing hands. Sunal watched sadly as it ascended to the cave mouth, which was hung with colourful flags to keep the spirits of the forest at bay.

The cave smelled strongly of decay and the floor was powdered with the dust of decomposed bones and wood. Along one side lay a huge stack of many hundreds of coffins with the associated débris of centuries of use. I gazed in wonder at the beautifully decorated knives and swords, intricately carved boat paddles and fabulous, early chinaware bowls and plates. It seemed incredible that the Dusuns, who lived in poverty, should keep such a valuable hoard of treasure around their dead. But these relics had become sacred and could never be removed. Ever-watchful Manoi Salong would see to that. I glanced curiously to the cleft in the roof, the home of Manoi Salong's head, the only part of him brought back to the cave when he was killed on a trip to the Kinabatangan River. The coffin of another hero lay on a large slab near the entrance. Carved in the shape of a buffalo's head it was painted in many colours.

Zinah's coffin was added to the tall stack. The parasol was placed over it and Sunal laid a bowl, some rice and water, a spoon and a *sarong* for the use of her spirit. An old man beat the coffin three times with a *rotan* cane and chanted an ancient prayer telling Zinah's ghost to remain content where she was and not to return to the *kampong* to haunt her family.

As I compared the recent, cheap, plastic plates and tin mugs with the wealthy gifts below, I felt I was witnessing not just the funeral of Zinah, Sunal's wife, but the death of a whole culture. The Dusuns were once a proud people with a good living from their fields, the forest and the river. They had hunted with blowpipes, spears, traps and hooks. They had taken the rhino for his horn, the elephant for his tusks and the crocodile for his valuable skin. In the coastal towns they had sold the meat of deer and pig and the fatty *tarpa* fish. From the forest they had collected *rotan* canes, aromatic *damar* resin, wild spices, sweet durian fruit and the strong red *belian*-wood. They had panned gold from the river gravel and their rice fields and fruit groves had been well cultivated and productive on the rich silt soil. Their jungleware was traded for the cloth and knives they needed and their only enemies were occasional headhunters or raiding pirates. The treasure in the cave and the lavish scale of their ceremonies suggested the culture of people only recently reduced to poverty. Now only by going into town could a man earn a regular wage. Changing times and the ravages of floods and cholera have depleted the population. There is no longer a market for *damar* or pig-meat and the hunting of other game or the selling of timber is now illegal. To-day's wealth lies in the rubber and oil palm estates or the large timber companies. The remote riverside *kampongs* have been left far behind by the mainstream of the new economy.

As we descended the long ladder, Bahat pointed out a smaller cave in the cliff face below us. A human skull grimaced among the bones and coffins that littered the floor.

'That is where the Ibans and Kayans may bury their dead. Only Dusuns may use the main cave,' he said proudly.

'So if I die you will put me in there?' I queried. Shocked, he replied, 'Oh no, tuan, you would be treated like a Dusun!'

We prepared to return to the *kampong*, where another two nights of feasting were necessary to fulfil the family's obligations to the dead. Before we left each member of the party picked a sprig of green leaves to wear in his clothing or hair. These would be discarded in a washing ceremony on our return but it was only fitting that we who had taken to Batu Balus something of the old life should bring back something of the new.

Chapter six

Hidden Lives

The cold, wet winter months passed slowly, but gradually the rainstorms grew less frequent and the river water became quite clear so that we could hunt for prawns among the gravel. Species by species the great trees came into flower, white, yellow and pink till the forest floor was littered with a soft confetti of petals. Ants worked furiously collecting the fallen blooms and dragging them off to their subterranean fungal gardens. Noisy swarms of bees flew overhead in search of safe branches from which to hang their rich yellow honeycombs. Dancing along the streams and river banks came an endless procession of gaudy butterflies; fast-flying swallowtails, electric-green birds-wings, lazily-gliding spotted milkweeds and a mass of whites and sulphurs. At puddles in the forest or on the sandy beaches they congregated in flickering carpets only to explode again into living snowstorms as I drew near.

The variety of insects and the subtlety of their devices was a constant source of pleasure to me, saving me from boredom during the long hours I spent alone in the forest. From odd moments of observation gathered over many months I was able to piece together a clear picture of their complex life histories and close inter-relationships. Metallic-winged blue butterflies bore shimmering, false 'eyes' and trailing, white tails to draw the attention of predators away from their vulnerable heads, but the hunters were no less crafty and many an attractive flower was not all that it seemed. White crab-spiders sat patiently on orchid petals ready to seize visiting butterflies but even more remarkable were the mantids. Coloured and arranged to look like flowers, they lured unwary insects into their waiting arms.

Fierce-jawed weaver ants searched through the leaf-litter for cockroaches and crickets, which they dragged off to their treetop nests, several ants co-operating to haul a heavy captive up the trunk and over the branches to the well-guarded entrance. At the slightest disturbance a hundred angry ants would swarm forth to surround their home in a protective wall of gaping mandibles. The nests are made from broad, green leaves bound firmly together with a web of silk, yet the adult ants have no way of producing thread themselves. Only the ant larvae have silk glands and the adult workers sew the leaves by holding a grub in their jaws and moving it back and forth like the shuttle of a loom.

Black and yellow mud-dauber wasps scoured the treetops attacking small spiders, which they paralysed with their stings. This was the beginning of a cycle in which the helpless victims are entombed in mud chambers on tree-trunks and the wasp lays a single egg in each before sealing the entrance. The developing grub munches its way through its living food supply, which remains fresh and does not rot in the tropical heat. When the larder is bare the grub pupates and eventually a new wasp breaks out of the chamber.

Sometimes, however, it is not a mud-dauber wasp that emerges but an intruder, a black, red and white velvet wasp. The velvet wasps play the cuckoo and lay eggs in the chambers made and provisioned by other species. Their grubs hatch quickly and immediately destroy the eggs of the rightful owners before fattening up on their poached caches of spiders. Male velvet wasps have wings but the females do not and hence look much like ants. They have such a painful sting and conspicuous colours that predators soon learn to avoid them. Other insects have taken advantage of this and wear the same bright warning colours so that they, too, are left alone. I collected a wide range of beetles, wasps, grass-hoppers, leaf-bugs, spiders and even moths all mimicking the gay velvet wasps and many had copied not only the exact colours and patterns but also the jerky, shuffling gait of their stinging models.

Perhaps the most amazing creatures I met were the forest gliders. Since most of the food, fruits, leaves and insects, is up in the canopy, there have always been great advantages for those animals who could travel easily from treetop to treetop without having to descend to the ground. The orang-utan does this in his own way, swaying trees to and fro until he can reach across a gap, but mainly it has been the small agile forms that have perfected this mode of life; animals with good balance and leaping ability, monkeys, squirrels, martens, civets and long-legged agamid lizards. Possession of a parachute or gliding membrane enormously increases the leaper's range as well as eliminating the danger of falling. The nocturnal flying squirrels have achieved this by the development of a broad web of skin stretching between the arms and legs. Stabilised by their long tails, these delicate animals can glide over a hundred yards. Often in the evening I have watched them come out to search for food. Occasionally one would fly the whole width of the Segama River to land on a far tree-trunk, bound awkwardly to the top and float off again through the forest. Twice I saw resting squirrels disturbed by inquisitive orang-utans and glide away to look for safer holes, their chestnut fur shining in the midday sunshine.

Another gliding mammal is the rare flying lemur, whom I saw only five times during the entire study. Like the squirrel he has a web of skin linking his arms and feet, but this also extends to include his tail and elongated fingers so that the whole animal resembles a triangular kite. Clinging to a tree-trunk, with his mottled, almost green, coat the lemur is invisible as a lichen-covered bump. Only his gleaming eyes and pink ears give him away and a knock on the tree will send him skimming off with incredible manœuvrability. The flying lemur is something of a zoological mystery as it seems unrelated to any other living animal. With the head of a mousedeer, the coat of a rabbit and the wings of a bat, he stands alone as one of Nature's greatest oddities.

It is not only among the mammals, however, that gliding has been evolved. Many of the tree-frogs bear wart-like suckers on their toes, enabling them to gain a firm grip on twigs and leaves as they land after leaping. One rare species, found by the great naturalist Alfred Russell Wallace during his explorations of the Malay Archipelago, even has elongated toes so that the extended, webbed feet act as tiny parachutes. Side flaps of skin and its flattened form enable one species of gecko to 'fly'. At rest the flaps fold under the belly like a waistcoat as the gecko clings flat against a tree-trunk. This weird creature can change the colour of its mottled skin or even the iris of its eye to merge into the bark. I do not think these animals were particularly uncommon but they were so inconspicuous that I found very few of them.

By far the most spectacular of the gliding reptiles was *Draco*, the flying dragon. There were several common species, the largest being over a foot in length. Their home was the vertical world of tree-trunks, where they climbed, seizing and devouring the ants that marched past. The males defended territories and signalled frantically to one another with evertible throat flags of bright red, yellow or black with white. Sometimes they engaged in wild chases, evicting unwelcome trespassers from their pitch. Females were more modestly attired with small blue flags on the neck and throat. Like the flying geckos they could change colour to suit their background, but whilst the females assumed inconspicuous browns and greys the males preferred bright green and yellow. The thin stick-like animals worked their way slowly up tree-trunks, then reaching the top would give a sudden turn of the head to choose another tree before launching themselves into the air. Suspended on broad, red, membranous wings, they glided down to a gentle landing, losing little height and showing great skill at turning corners or even executing complete loops in mid-air. The wings are attached to ribs which project through the body-wall and can be erected or folded flat to the side under muscular control. For travel and feeding the animals are quite independent of the ground but the female must visit the forest floor to lay her eggs. After a shower of rain she descends to dig a small hole, deposits three or four eggs, hastily hammers down the soil again with her head then hurries back to the safety of the trees.

The golden flying snake swims sinuously through the air, flattening its body to increase the gliding surface. These snakes flow through the branches at amazing speed and I have seen them leap across gaps of over fifteen yards. One specimen I found was held in a deadly embrace with a large gecko. The snake was firmly twisted around the lizard but before it had died the gecko had bitten its attacker and clamped on to his mouth. Rigor mortis had trapped the snake so that he could neither flee nor eat his prey. I eased the dead animal's jaws open with my knife, thus freeing the snake. Paying no attention to me he proceeded to devour his prize, although it was a good deal larger than himself. This Herculean task accomplished, he skimmed easily away and up the nearest tree, quite unhindered by his extra load.

Snakes are animals that have always fascinated me and they were very plentiful in the forest. Since they are not conspicuous I generally overlooked them, but whenever I did spot one I usually saw others on the same day because my mind and eyes had become attuned to the right 'search image'. There are thirty poisonous varieties in Borneo but these are mostly tree-living, and one is probably less likely to be bitten walking through the forests of south-east Asia than anywhere in Africa, India or Australia. Only the yellow and black banded kraits and the king cobras, which reach over twelve feet in length, are a real threat to man. Cobras were usually quick to get out of my way but on one occasion I heard a rustle and looked down to see one strike at me. I jumped aside as he aimed and he fell a few inches short. He reared and struck again but by this time I was well out of range. I caught another large cobra with the aid of a stick and took it back to camp, where I let it bite my hat several times to empty its poison sacs before releasing it. I obtained some spectacular film of the angry creature rushing forward with its broad hood extended as it threatened the camera.

The reticulated pythons that lived by the river were enormous and we saw several over twenty feet in length. Their young could be found curled up in trees overhanging the muddy waters whenever a flood had obliged them to leave their dark hollows. I kept a pair of these snakes in camp in a special cage and fed them on the rats that invaded our houses. My men refused to believe they were not poisonous even when I myself was bitten without any ill effects. They believed I had magic against such venom so I was not a fair test but none of them was willing to try.

Bahat once found the carcass of a fully grown pig which had been deserted by a python. The pig was curiously striped in blue and green bruises, where it had been squeezed to death by the powerful snake. A few days later we were out hunting when we heard barks and screams from one of the dogs. We rushed on to find the poor animal lying quivering with shock while a large python twisted sinuously round her. Pingas leaped forward and slashed the snake's tail. Instantly the python released its victim and fled for the river. Bahat was furious and set off to take revenge on the snake. Remarkably enough he found it again, hiding under the bank of the river, and speared it savagely to death. Paying no heed to my request to save the skin he stabbed it again and again till it was a bloodied mess. The bitch had a huge bleeding bruise on her neck where the python had struck her but recovered surprisingly quickly.

One of the other dogs was bitten by a small poisonous snake in camp and was very ill for a couple of days with fever before she, too, recovered. Three more of our canine companions disappeared without trace, but to compensate for their loss two of the bitches went off into the forest and dug deep dens, where they gave birth to a total of nine pups. These were greatly loved, though not at all well treated, by the children.

On my travels one day I passed a large tree with a yawning hole at its base. Loud grunting noises emanated from within. As I watched, the head of a porcupine appeared. On seeing me the frightened beast raised its armoury of spines and tried to back away down the hole but the erect quills prevented this manœuvre. He had no alternative but to come forward, so snorting furiously the animal charged out, circled the tree at a gallop and vanished head first down his hole again. It was a ludicrous spectacle.

Another animal that could give a newcomer to the forest quite a fright is the muntjak. This little red deer looks harmless enough but whenever he is disturbed he lets out such a terrifying bark or roar that it sounds as though a tiger, at least, must be lurking in the undergrowth. The male uses this bark to defend his territory and the Dusuns have learned to imitate the noise by blowing between two leaves. In this way they can lure either the little stag to challenge the intruder or a doe looking for a mate to their calls.

The largest deer in the forest is the sambur, which stands some three feet high at the shoulder. In the daytime these animals hide in the thickets and are rarely seen but at night they emerge to graze on the grass and ferns along the river banks. I had a permit to kill one deer a month to supplement our pot and Bahat would go out hunting for them after dark. Drifting silently down-river, he scanned the banks with a small paraffin lamp until he saw the eyes of his quarry glow dull red.

At the other end of the scale is the tiny mousedeer; unlike his big cousin, he stays in the forest feeding on fallen fruit. When I walked through the night I often caught this timid creature in the light of my torch. Dazzled, he would freeze, gazing back with large frightened eyes.

Other animals active at night included the slinky striped civets, bushy-tailed *binturongs* and amazing scaly pangolins. Pangolins live on the eggs of ants and are such agile climbers that they can raid treetop nests as well as those in the ground. With his powerful front legs the anteater rips open a nest while his long sticky tongue flashes in and out to collect the eggs before he has to make a hasty retreat from the attacking swarms of biting ants. I tried to keep a mother and her baby as pets. The baby was very sweet, riding in style on his mother's tail as she travelled. Unfortunately, they forced their container open in the night and made a moonlight getaway. Another young pangolin was brought to me and kept me busy cutting down ants' nests for his meals. When he became sickly and ignored his food I decided to release him, too. Happily he returned to wild living and wandered into camp a few weeks later looking very fit and well.

Another curious night animal is the moonrat. This large slow-moving shrew has such a fearful smell that no-one will attack it. With nothing to fear, the moonrat advertises himself in the most conspicuous manner. His white form stands out like a creeping ghost as he wanders the dark forest floor, waving his long nose from side to side sniffing out his beetle dinner.

The call of *Kawau*, the argus pheasant, is one of the most familiar sounds of the forests of South-East Asia yet this bird is rarely seen. In 1869 Wallace reported, 'This was the country of the great argus pheasant, and we continually heard its cry. On asking the old Malay to try and shoot one for me, he told me that although he had been for twenty years shooting birds in these forests he had never yet shot one, and had never even seen one except after it had been caught.' At night the pheasants roost among the low branches but they spend all day on the ground. The cock clears away fallen leaves and seedlings to make a bare, circular dancing area about seven yards across. Some of these dancing rings have been used for hundreds of years and the ground has become packed hard. From eight o'clock to noon each morning the male sits on his ring calling periodically. The bird makes such a predictable and vulnerable target for any stalking cat or civet that it has had to develop the sharpest sense of hearing and eyesight. Often I have crept cautiously up on a ring to find the bird already gone or catch only a fleeting glimpse of him running down one of his many escape routes into the undergrowth. Nor was it possible to make a hide overlooking such a ring as the birds were very sensitive to any change in their surroundings and would desert a ring where I had been tampering. Only twice did I see the cock calling from his ring, giving a double bob of his head in time with the clear two-toned hoot. The birds called every eight seconds or so for several minutes, then sat resting or paraded the ring clearing new-fallen leaves, before starting on another session. Away from their rings the pheasants wandered about the forest, also calling but in a distinct series with some thirty single hoots in quick succession.

It is in his display dance towards the hen that the cock pheasant reveals his most spectacular tail and long wing feathers. On one lucky occasion I was able to witness this unusual sight because the male was too engrossed in his courtship to notice my approach. As I came towards his ring I heard a curious rattling noise. I crept forward carefully till I could see the bird strutting to and fro in the middle of the dance floor, pecking rhythmically at the ground with his tail held high. Suddenly he fanned out his long wing feathers into a broad shield in front of his head. Like a peacock's tail the fanned plumes were bedecked with beautiful eye-spots to dazzle his impressionable mate. He held his wings fully extended for several seconds then started to fold them, slowly shaking the feathers back and forth to produce the strange rattling sound that had first attracted my attention. Three more times he repeated the bizarre performance then stopped abruptly. Perhaps his lady-friend had seen me and fled, for suddenly he stretched his blue head high, glanced about him and scurried off down the path.

The forest was home for many other interesting birds; gaily coloured pittas and brown babblers hopping along the ground, sweet-singing drongos and dazzling bee-eaters in the canopy. Many had curious nesting habits. Tailorbirds constructed leaf nests, industriously sewing together leaves with thread collected from spiders' webs, while the swiftlets and frogmouths had abandoned nests altogether and cemented their eggs on to slender branches. Among the most spectacular of all were the hornbills, for not only were these large, noisy and fantastically decorated birds, their nesting behaviour was among the most unusual. Having selected a suitable tree-hole the male hornbill seals up the entrance, trapping his mate inside and leaving only a small slit through which he passes her fruit each day. Within her chamber she lines the nest with her own feathers and lays her clutch. The eggs are laid several days apart and incubated continuously so that the first egg laid is the first to hatch. When all the young have emerged the female breaks out of her prison and reseals the entrance. The parents feed their brood of growing squabs through the hole and as each youngster fledges he in turn breaks out, but always the wall is repaired until the last young has flown. There were many species of hornbills, some gregarious like the black hornbill and the noisy pied hornbill, but most of the larger types were solitary. The helmeted hornbill has an ivory casque on his bill which is so much prized by Chinese carvers that in populated areas the bird has become quite rare. The rhino hornbill, on the other hand, takes his name from the incredible curved horn on top of his head. This species is much revered as a sacred bird by the Iban Dyaks, who make elaborate sculptures and perform exciting dances in his honour.

Birds are important to both the Dyaks and the Dusuns as messengers of the guiding spirits and a great science of omen-reading has been developed to enable correct interpretation of their messages. I wasted several infuriating days when the men refused to work because they had heard the alarm call of the shy banded kingfisher or a crimson bee-eater. To go hunting after such a warning would be to invite certain misfortune or even death.

Another animal regarded with great superstition was the tiny tarsier. I was very keen to find some of these big-eyed rarities but not only did I fail, I could not get any help in looking for them because of the bad luck that befalls anyone who sees such an animal.

One curious beast of which there was no shortage, however, was the extraordinary proboscis monkey. Around the coast these large ungainly creatures are well-known inhabitants of the mangrove swamps, but they also live deep in the Ulu Segama forests and there were several enormous troops with up to a hundred animals in each. Because of their large red bulbous noses these monkeys are locally known as *kera belanda* (Dutchman monkey) and the old males do look simply incredible with their ridiculous snouts and close crew-cuts. At night the spread-out groups keep in contact with obscene nasal honkings. Probably the animals' noses help to produce this unusual tone but they are also used as a powerful visual signal and are presented as a threatening gesture to both man and monkey alike. These monkeys were far from friendly and whenever they saw me would rush overhead releasing a flood of powerful-smelling urine. Indeed, so strong was their odour that I could find the animals from some distance away by following my nose. When travelling the troops made such a noise that sometimes I thought a whole herd of elephants must be crashing towards me. One of the most interesting talents of the proboscis monkey is its ability to swim. It is not the only monkey that will happily take to the water – I have seen the long-tailed macaques swim across rivers – but the feats of the proboscis monkey outdo those of his relatives. Several times when we disturbed them feeding in riverside trees they made spectacular leaps into the water and vanished completely. Perhaps a large noseful of air enables them to stay longer under water; certainly such ability is a great asset in their swamp and riverine habitat. Still little is known about these fascinating animals and I am sure a more detailed study would prove rewarding.

Although bears were quite common in the forest they were generally shy and I rarely saw them. At my first encounter it took me some time to realise what I had come across. I could see two low black forms running about in the undergrowth, bubbling like a steam engine. I expected a couple of turkeys to appear but when the playing animals came into the open I saw that they were two sun-bears. One was large, about a hundred pounds, and the other was about half the size. I decided they must be a female and her cub. They chased each other madly round the base of a tree in a fast lolloping gait, then suddenly turned and were running towards me. The female in the lead did not notice me until she was already too close to swerve. She roared and charged straight on. I drew my *parang* and swiped at her head as she reared up on her hind legs. She avoided the blow and I swung aside to miss her sharp claws. Her momentum carried her on beyond me but I lost my balance and fell. The cub chose this moment to rush in to the attack. I swung my blade, striking his shoulder. He fell back howling and I turned just in time to see his mother coming at me for a second rush. As she drew near I let out a full-blooded bellow and she veered away to the side. Seizing my chance I scrambled up the hill, glancing back anxiously to see if she was following. She was running to and fro around the injured cub, barking and roaring, and I hurried on, feeling very sick as reason took over again from instinct. The mother bear grabbed the howling cub by the neck and walking backwards hauled him away across a small stream and up the opposite hillside.

I had to sit down for several minutes before my limbs stopped trembling and I could set off for home. I remembered Meng telling me that the bear was the most dangerous animal in the forest because his eyesight was so bad. While most animals will go out of their way to avoid men, the bear blunders into fearful proximity, suddenly realises his danger and panics, lashing out furiously with his lethal claws. All things considered I was lucky to have emerged unscathed. Bahat, my boatman, had a permanently stiff foot as a result of an encounter with a bear and others, less fortunate, had even been killed. In future whenever I met these animals I always equipped myself with a stout sapling and waved it frantically while shouting at the top of my voice to warn them of my presence and give them time to avoid me.

The only other animals which really scared me were elephants. I avoided going to places when I knew they were present and loud crashings in the undergrowth would send me scurrying off on wide detours. Down by a stream near one of my forest shelters I kept two drinking tins hanging in a tree. One morning when I passed the cups were missing and all around were the tracks of a large elephant. About fifty yards away I saw something shining in the water and fished out one of my tins, now crushed flat. Nearby I found its mate, which had suffered similar treatment. I don't know whether it was their shiny appearance or the smell of man that had attracted the elephant's attention but he left me in no doubt about his feelings on litter in his domain.

After several months of avoiding the great beasts, or catching only brief terrifying glimpses of them as I retreated, I decided that I really must take myself in hand and try to get close enough to the elephants to film them. After a disturbed night in the forest listening to trumpetings and noisy crashing down the hill, I set off with my cine camera to follow the animals' tracks. Large piles of droppings steamed menacingly in their wake and the paths crossed back and forth indicating that several beasts had passed this way. At last I could hear them ahead making a curious banging noise, not quite regular. My knees were shaking and I was itching to be away but I forced myself on till I had located the elephants in the stream bed below. I glanced from side to side as I approached, noting the position of any climbable trees in case of an emergency. Cautiously I advanced to the edge of the rise and peered through a bush. Standing in the stream with her back towards me was a large cow and in front of her a smaller animal feeding on vines. The female pulled the fruiting creepers down with her trunk and stuffed them into her mouth. She nodded her head from side to side as she ate and shifted her weight to another foot, knocking the boulders together to produce the strange noise I had heard. I tried to hold the camera steady but my hands were trembling so much that I had to rest it on a branch. I shot off a few feet of film then hurried away in case the elephants were disturbed by the sound. Only when I was half a mile away did my heart stop its violent beating and I slowed my pace.

Another animal reputed to be dangerous was the *Banteng* or wild cow. This animal is rather rare and heavily protected by game laws. It is the size of a buffalo and the bull looks very striking with his black coat and white rump and socks. The cows are smaller and warm chestnut in colour. The tracks of these animals are very similar to those of domestic cattle but uncommon so far into the jungle. I saw the animals only three times and always they were inquisitive, standing watching me as I approached, then cantering ahead to stop and stare again. One day we found flattened grass beside the river, and human footprints and patches of blood informed us that poachers had been at work. Hidden in the ferns were the discarded head and guts of a large bull *Banteng*. Bahat explained that the meat would fetch a high price in the town as beef.

One of the rarest animals of the forest was the two-horned rhino, much prized for its fabled horns which are supposed to have great healing properties and act as a powerful aphrodisiac. At the turn of the century the rhino was not uncommon in the Segama area and they were often hunted by Dusuns with blowpipes and spears but the advent of the shotgun seems to have been the animal's downfall. They have become so scarce that few Dusuns to-day have ever seen the tracks, let alone a rhino in the flesh. Ibans from Sarawak still make a living from poaching the last few remaining rhinos and would spend weeks at a time tracking the beasts. Although I had seen fresh rhino tracks on my third day in the jungle I found them again only a dozen times in sixteen full months in the Ulu Segama. They were usually around the mud wallows in the hills at the north end of my research area. I never had a clear view of a rhino, though once disturbed a large animal who crashed away down the slope, and, despite all my subsequent efforts, this was as close as I got to the rare creature. On another occasion I followed two sets of fresh tracks for over an hour in the hope of seeing the beasts. A strong animal smell still lingered so I knew they could not be far ahead. I could see where they had trampled down small bushes to feed on the leaves and the tree-trunks were muddy from their passing. Unfortunately the tracks came out on to hard gravel and although I cast far to each side I found no further trace and had to give up the trail.

The rhino may be rare but at least it is a well-known and scientifically documented animal, which is more than can be said of *Batūtūt*. I was travelling alone along a hill ridge on the far side of the river where I had never ventured before. The path was good, though rather muddy, and I hadn't a care in the world. Suddenly I stopped dead, amazed at what I saw. I knelt down to examine the disturbing footprint in the earth, a print so like a man's yet so definitely not a man's that my skin crept and I felt a strong desire to head home. The print was roughly triangular in shape, about six inches long by four across. The toes looked quite human, as did the shapely heel, but the sole was both too short and too broad to be that of a man and the big toe was on the opposite side to what seemed to be the arch of the foot.

Farther ahead I saw more tracks and went to examine them. There were imprints of both left and right feet, though which was which I could not tell from their curious distribution. Many of the prints had been obliterated by recent pigs but a few were quite clear and I made drawings of some of these and notes of their relative positions. I found two dozen footprints in all, scattered along some fifty yards of path. Still uneasy about my find I continued along the ridge until I reached terrain I knew quite well.

Perhaps my mind was preoccupied with other things for I could find no sign of orang-utans. I was quite happy to abandon my quest and shelter under a leaning tree-trunk waiting out a sudden rainstorm. Thoughtfully I made my way back through the dripping forest to the river, where Bahat was waiting with the boat.

Back at camp I showed him my sketches and asked what animal could make such tracks. Without a moment's hesitation he replied '*Batūtūt*' but when I asked him to describe the beast he said it was not an animal but a type of ghost. Bahat gave an imitation of its plaintive call, a drawn-out *tootootootootoo*, from which it derives its name, and told me many stories about this shy, nocturnal creature, who lives deep in the jungle feeding on river snails, which it breaks open with stones. *Batūtūt*, he told me, is about four feet tall, walks upright like a man and has a long black mane. It is said to be fond of children, whom it lures away from their villages but does them no harm. To adults, however, it never shows itself, but occasionally men had been found that *Batūtūt* had killed and ripped open to feast on their liver (to Malays the seat of all emotions, analogous to the European heart). Like the other spirits of the forest the creature is very shy of light and fire. Bahat said that, as a young boy, he, too, had seen the footprints of *Batūtūt* and other villagers had also seen them from time to time.

When I suggested that perhaps a bear could make such prints, his pride was wounded.

'They are too big for a bear and they have no claws. Moreover, bear tracks are a different shape.' Looking down at his stiff left foot, he added, 'I should know, I have had plenty to do with bears, tuan.'

When I made further inquiries in the *kampong* I found that *Batūtūt* was quite well known and other stories confirmed what Bahat had told me. I secured photographs of the feet of sun-bears and indeed they were too small and differently shaped to be responsible for the tracks I had seen. Later I saw plaster casts of even larger footprints from Malaya that had definitely been made by the same animal, there known as *orang-pendek*, or 'short fellow'. Again, natives spoke of a small creature with long hair, who walks upright like a man. Drawings and even photographs of similar footprints found in Sumatra are attributed to the *Sedapa* or *Umang*, a small, shy, long-haired, bipedal being living deep in the forest. An interesting feature of these accounts is that the feet of the *Umang* are said to be back to front; an idea which stems, I think, from the position of the big toe on the outside of the foot. Among bears the largest toe is usually on the outside and the foot-pad is broad and triangular. It would, however, take a much bigger bear than a sun-bear to produce such prints. Could the forests of South-East Asia conceal a still-undiscovered species of bear, or are we to believe the tales of the mollusc-eating little-folk?

Chapter seven

Fruit Harvest

During March and April the wild pigs drifted back but in a very different condition from the sleek, healthy animals that had headed north a few months earlier. They were so thin that some were dropping from starvation and they were much fiercer now so that I had to be wary in case they should attack me. Pingas speared a pig not far from camp but there was so little meat on the scrawny creature that it wasn't worth carrying it home. By the next day there was not a trace of the corpse. Its ravenous colleagues had torn it to pieces, carrying off bones, teeth and all. They were so desperate for food that the swine took to waiting under fruit trees to gobble up any scraps dropped by feeding monkeys or squirrels. Fierce fights broke out among the hungry pigs but the diners took no notice. Even the orang-utans paid no attention to these noisy table mates and I was rather hurt that they were so tolerant of other rowdier intruders when they became upset and indignant at my quiet presence. Perhaps I should join in the mêlée squabbling over fallen husks!

Suddenly in June the population became more active. The males began calling vigorously and many unfamiliar orangs arrived from the west. Sarah, the lone female, was the first to meet the invaders. She squeaked excitedly and shook branches at a strange female and juvenile passing through the trees below her. Half an hour later another female with two young appeared, closely followed by two large males displaying and branch-shaking as they progressed. The red sub-adult was obviously fed up at being pursued by his angry elder and took out his ill-feeling on Sarah, who scurried away protesting at the injustice. The big, black male decided that I would make a better victim. There were eight orangs in the neighbourhood, a wonderful opportunity for observation, but when the menacing ape climbed to the ground to meet me I decided it was time to leave.

Within a mile of this unfriendly crowd I spotted a familiar brown shape clinging to the white bark of a *Polyalthia* trunk. Midge was a few feet below his mother and both were happily engaged in chewing strips of bark, sucking out the sweet sap and spitting out the remaining wodges, just like old men chewing on their tobacco. They seemed perfectly at ease but cannot have remained unaware of the presence of other orangs for very long for the next three days were frequently disturbed by the booming calls of the aggressive black male. Margaret and Midge meandered slowly towards their old haunts on Horseshoe Ridge. Eventually I became bored with their shyness and sluggish pace so headed north again to see what the newcomers were up to. I had another amazingly full day for they were still moving in a tight-knit group and I was able to watch nine different animals. Most were timid females accompanied by their young but a sub-adult male, who I felt must be Humphrey of the winter before, was very tame and let me sit close to him while he enjoyed a meal.

Day after day I followed different members of this band, which was rolling slowly eastwards. Individuals wandered wide to north and south but there seemed to be a definite group nucleus where perhaps half the emigrants would be congregated at any one time. This was quite different from anything I had seen before in an entire year's fieldwork. I caught one more glimpse of my black antagonist storming past along the ground, too intent on his mission to notice either me or the two orang-utans swinging above. After that I saw no more of him but the other travellers continued steadily and crossed Central Ridge into Redbeard's domain. Redbeard seemed to have been expecting them for I had heard him calling several times during the preceding week. He certainly gave them quite a welcome, hurtling through the trees to attack two helpless females and their young. They fled screaming up a narrow valley but Redbeard chased after them remorselessly. Catching up with the smaller female he dragged the unfortunate miscreant out of the tree and beat and raped her before moving on up the ridge, calling as he went.

Ironically in this time of famine for the bearded pigs, the sweet fruits they would have welcomed were gorged on by orang-utans and monkeys before they had a chance to fall. There were even few of the less popular, oily dipterocarp seeds which maintained the pigs at other times of the year.

The climate in tropical rainforest remains very similar throughout the year. Although in the dry season there is only half as much rain as during the wetter months, the dark jungle is always moist and the temperature stays much the same, rising to about 90° F. in the heat of midday. It is so damp that leather quickly perishes and camera lenses support flourishing growths of fungi. Nevertheless, in spite of the year-round uniformity of heat and humidity, definite seasons do prevail and are no less obvious than those of the temperate regions.

Many of the trees have their own appointed times for flowering, fruiting and putting out new leaves so that the commoner species can greatly affect the flavour of the whole forest. When the *Melapi* trees are in bloom whole hillsides are white with their blossom and the *Ramus* vines decked in new leaves tinge the forest with red. Since most seeds ripen between April and November there is a recognisable fruit season comparable to a temperate summer. This time of plenty is heralded by the *Wadan*, the tall, climbing bamboos, whose hard, nutty produce are a common and important item in the diet of the orangs. As the months pass the wild plums, lychees, *rambutans*, taraps, *lansats*, figs and durians all ripen to yield a generous harvest. But with the approach of another rainy season the wealth and variety of fruits starts to drop off so that the animals must make the most of the lavish crop while it lasts.

Throughout the fruit harvest the orang-utans feed busily, stocking up on the sweet foods and building up great stores of fat to last them through the coming winter when they will be reduced again to a meagre diet of leaves, bark and wood pith. It is because of the seasonal variation in food availability that orangs have such a capacity for fat storage, but it is this same safety mechanism which causes captive orang-utans to become so hopelessly obese when they are offered a diet rich in fruit all year round.

Although the fruit season is in many ways a time of plenty the harvest is unevenly distributed. One part of the forest may boast a good crop of *rambutans*, while another produces an abundance of figs. Moreover, not only does each species ripen in its own time but trees of the same kind, but in different localities, bear fruit out of synchrony. To gain the most advantage from the rich crop animals must follow the harvest so that they are in the right place at the right time. Hence this is the season when orang-utans are at their most mobile wandering on long forages throughout their range.

At the beginning of the season the orangs were not travelling far and I was able to keep a close watch on several of the residents in my study area. The extreme eastern end of my range was occupied by two females, one accompanied by a juvenile and the other with a juvenile and an infant. Although they did not travel around together I often found them feeding in the same tree and suspected them of being related. Perhaps the females were sisters or even mother and daughter. Margaret and Midge were still living at the far end of my territory but they were an elusive couple and I saw them only rarely. Another lone female lurked in the hinterland and there were several males present but these were less predictable in their movements. Redbeard held court to the north-east and Raymond visited us occasionally but of Harold there was no sign. Other patriarchs made brief appearances but all continued ponderously on their way.

At the head of the valley stood a large tree of bitter wild mangosteens and alongside a laden fig, its succulent fruit not yet ripe. As this hill seemed to lie in the direct path of the migration I decided that rather than roam all over the country in search of orangs I would station myself at this feeding site and see who passed through. This decision proved to be very profitable for it involved much less work, and twelve different orangs came to feed there during the next seven days. When the last stragglers had passed through the gap I circled round the advancing group and caught up with the vanguard again about a mile farther east. Redbeard had taken over the central position as co-ordinator of the party and wherever I heard him calling I could be sure to find subjects to watch.

Gradually the pilgrimage turned northwards as the strangling figs began to ripen. Fig plants are really climbers using other trees for purchase as they twist down from the canopy and smothering and killing their supports as they increase in size and spread their leafy branches. Some strangling figs have many aerial roots fanning out towards the ground like flying buttresses but the species now attracting attention threw out only a solitary white root. Thus the thick trunk supporting the heavy crown of figs was merely a dead prop and the lifeline to all the greenery was a single, thin strand hanging down like any liana. One such tree bore

The endless green jungle seen from Lookout Peak and, below, the camp with its grass roof at Samparan

The orang is a rather solitary ape. Below, I wondered if I wasn't becoming half a mawas myself

The villainous Humphrey bides his time and, below, appears with Ruby and Richard

a. Red crested partridge
b. Slow loris
c. Rajah Brooke birdswing
d. Frilled lizard
e. Green Tree Snake
f. Spiny tortoise

a fig root extending well over a hundred feet which reached the ground yards away from the base of the leaning trunk. This tree was set apart from its tall neighbours so that the only route the orangs could use to reach the tempting fruit was up the swaying vine. It took an orang-utan about three minutes to climb this fine rope and almost as long to descend again. The situation was ideal for some spectacular photography and I cursed my cine camera, which was limited to a fifteen second run.

More amazing sights were in store. When a nearby fig ripened a few days later I was able to break all records for orang watching. My red friends had found the rich crop before me and there were already four large, leafy nests high among the branches. The tree was still thickly covered in figs, however, and there were no orang-utans in sight. It was midday and stiflingly hot. Then I spotted a tuft of bright red hair poking out above the brim of the lowest nest. For an hour nothing happened and then a chin appeared, a chin so broad and distinctive that there was no doubting its owner. Faithful old Redbeard was right at the centre of things again. He took his time to emerge but gradually his magnificent form hove into view. Hanging by one arm he swung leisurely out on to a peripheral bough as though his two hundred pounds' weight was of no consequence. Sampling only the ripest fruit he worked his way steadily along the branch. Contentedly full, he retired to his couch for a well-earned rest. So the day passed in an interminable series of tasty snacks and short naps. In the cool evening a flock of pigeons joined him, fluttering and cooing as they tucked into the ample feast. As the sky darkened the birds departed and I expected Redbeard to wander off and build a night nest in a smaller tree. But instead of descending he inspected his old sleeping quarters and finally settled for the accommodation he had used throughout the day.

By first light the huge tree was alive with feeding animals. Redbeard sat enthroned among the branches and eyed his guests in a lordly fashion. The pigeons had returned with two black giant squirrels and several cackling hornbills. At a respectful distance, harvesting the back of the tree, were another two orang-utans, a mother and small infant. The youngster was getting plenty of experience at fending for himself for he was quite unable to keep up with the speed at which his mother moved about selecting choice titbits. The twittering and wing-clapping of the doves attracted more of their friends and a trio of hooting gibbons leaped across the wide gap to join the throng. Using branches as springboards they rocked up and down before launching themselves into the air. The gibbons worked swiftly, plucking off the red figs and popping them into mouths that were already full. Deserting his elders, a young gibbon tried to persuade the infant orang to join in a game of poke and run, 'you can't catch me', but in spite of his equal size the baby was not up to such agility and scuttled back to his mother for protection. After a quick circuit of the whole tree the gibbons moved on again, leaping away on their rounds.

Another female orang-utan arrived with her juvenile and clambered up the fig root ladder. The first mother decided that the place was becoming too popular for comfort and prepared to leave. Arm over arm she cautiously descended the liana, looking all around to check that it was safe. With complete disrespect for his mother's careful precautions the baby came tumbling after her and clinging to the thin root slid down in one swift descent like a practised fireman.

As the day warmed up the doves and hornbills flew off and the other animals gradually dispersed till Redbeard was left in sole possession. He had moved his bed and was reclining in another of his four nests. During the afternoon more visitors arrived for a meal and he was soon entertaining a female orang-utan and her two young and a host of clattering hornbills. An adolescent male orang began to heave himself up the narrow rope but Redbeard uttered a string of deep bubbles warning the newcomer to keep clear.

Not daring to disobey, the poor youngster had to search elsewhere for something to eat. With evening the crowd thinned and the mother and baby returned for a quick supper.

I could hardly believe my good fortune. To be able to sit and watch so many orang-utans in action was almost unbelievable. For a year I had struggled through terrible terrain to gain a few hours' observation here and a brief glimpse there. With luck I might be able to follow an animal for three or four consecutive days but it had taken months to accumulate my data and then on only one or two subjects at a time. Now all I had to do was sit in the same spot and scribble continuously. It was so easy. In one week I spent over eighty hours watching orang-utans and that month I saw more encounters between different sub-groups than in the whole of the preceding year. Even now the orangs paid virtually no attention to one another but behaved as though they were feasting in splendid isolation. They were obviously maintaining close contact within the group, however, and this was completely unlike my earlier findings. How would it be possible to generalise about the orang-utans' way of life when they showed such contrasts of behaviour and who could tell what other facets of their characters still remained hidden?

Probably the most likely explanation of my contradictory discoveries lay in the fact that the region to the north of the Segama had been disturbed by timber-felling operations some twenty miles distant. Normally this area was occupied by a number of well-spaced residents but there had been an influx of foreign orang-utans in search of new homes after their old haunts had been destroyed by the felling activities. This would certainly explain why so many animals had appeared briefly in my territory but had never returned. The presence of these strangers had given me a rather confused picture of orangs' normal ranging behaviour. They also seemed to be so upsetting the resident population that the adult males had increased their rate of calling and the birthrate had dropped drastically.

I had missed the durian season the previous year but in July the prickly fruit began to swell once more. Durian trees are rare and I knew of only eighteen in my study area. They were of three varieties: *Lampoon*, oblong and green with creamy arils; *Didingi*, an uncommon round, red form with very sweet, scented, yellow flesh; and *Meraan*, a round, yellow fruit with a savoury red centre. The latter is a great favourite with the Dusuns but it was *Didingi* and *Lampoon* which were highly prized by the Chinese and fetched a good price in the local markets. Bahat was busy mending his honey-collecting tackle for he employed the same methods to gather durians.

For me the season started with our first sighting of a big, young male orang feeding high in a *Lampoon* tree on the southern bank of the river. We pulled our boat into the shallows and watched the feasting gourmet for an hour but he was so intent on the rich flesh that he never once raised his head to acknowledge our presence. I had forbidden Bahat to harvest the trees visible from the Bole and Segama Rivers or within my study boundaries. I visited them regularly to see if orangs had found them. A feeding ape discards the prickly shells which fall to the ground and gradually turn brown, and then black, so that it is possible to estimate from their condition how many days have passed since the animal's meal.

A week of bad weather killed off many of the young fruit so we did not get the bumper crop that Bahat was hoping for but the trees did give me a good indication of the relative numbers of orangs residing in different parts of the forest. To the north of the Segama many of the durians were raided five or six times and the fruit was never allowed to ripen, but to the south trees were only visited once or twice by orangs and the crop reached maturity. East of the River

Bole most of the durians remained untouched, a sure sign that there were few orang-utans in the area for I have yet to meet a red ape who can resist such tempting bait: no wonder I had never had much luck there. Within my northern region I cleared the vegetation round some of the more popular sites so that I could easily watch the feeding animals. The orangs had a good knowledge of the geography of the trees and I followed Humphrey on one well-planned trip that managed to take in at least half the durians in the neighbourhood.

Tell-tale fresh *Meraan* shells were scattered at the eastern end of my range but I could not find the orang responsible. Next day, however, I met the female Joel and her two offspring heading along the ridge towards the fruiting tree. Just as I had decided that these were the guilty parties two more orangs approached, Rita and her juvenile son, Roy. Rita kept her distance but Roy hurried across to play with the other youngsters. Quietly Rita slipped away down the valley and she was soon out of sight. When Roy noticed that she was missing he left his playmates and dashed off down the slope in search of his mother, squealing plaintively as he went. I followed the remaining trio up to the *Meraan*. Sure enough the floor was littered with new shells and Joel's family had to be content with Rita's leavings. It was not surprising that Rita had been so embarrassed at their encounter.

When the durians were finished a good crop of leathery, green pods on the hard, white *Merbau* trees kept the orangs busy. Many other fruits were also plentiful and the apes could laze or gorge as they desired. With the widespread abundance of food the population had become dispersed once more and I was limited to observing solitary males or females with their young. Redbeard had vanished and a strange male, Alex, wandered unchallenged over his domain. Margaret was still about and Joel stayed in the same valley. Since there was no problem in finding food the animals were able to return to their own favourite corners of the woods and enjoy undisturbed the easy-going, tension-free hermit life that is so characteristic of the species.

Only in times of local shortage would the orangs leave their preferred haunts and wander far and wide to find a more productive area. Only then would they follow the calling males, the animals with the greatest experience and knowledge of the forest and past seasons. If anyone could find food it would be the bellowing males but it was as well not to get too close to these bad-tempered, old patriarchs who resented the crowds they attracted.

Close to camp I bumped into a little orang female. Several juveniles had been alone but they were usually the more independent males. Perhaps her mother had died or suffered some accident, or maybe she was just particularly precocious. Near the top of the ridge she spotted Rita and Roy and headed towards them. She was shy of Rita but accepted Roy's invitation to play and they were soon involved in a game of rough and tumble among the hanging lianas. Back and forth they chased and swung, mingling orange with chocolate, while mother sprawled at ease, high on a leafy bough. For three days the newcomer stayed close to Rita and Roy, playing or joining them in a feast of rich, red taraps. Then Rita and son moved on and the little female was left alone to compete with the gibbons for the few remaining fruit. She was the last wild orang I saw in Borneo for next day we loaded the boats for our final trip down-river.

For the last time I floated past the familiar islands, hills and meandering streams that I had come to love. A few miles above the Tapadong caves we met the fruit bats that had been such a highlight of my first trip on the Segama. Disturbed by the boats, they poured off the laden branches of their roost. Thousands of leathery wings flapped in asynchronous rhythm lifting the shrieking bodies ever higher in tight spirals. We drifted on steadily, carried along by the fast-flowing current and after two miles or so I looked back to witness a memorable sight. The sun was setting in a blaze of crimson between the dark forest and the rising mountains and the sky was filled with bats. As far as the eye could see they were stretched out in a long, wavering ribbon that followed every bend of the winding Segama. The leaders were almost overhead as though escorting us in a final farewell before the cloud broke up. In twos and threes, nines and tens, they split off in small squadrons in quest of any young leaves and sweet fruits that the black, shadowy jungle had to offer.

Chapter eight

Interlude for Gibbons

The orang-utan is not the only ape in South-East Asia. He shares the vast tracts of jungle with his cousins, the gibbons and siamangs. The melodious chorusing of the graceful gibbons had much enlivened the Bornean forest and during my stay I had learned a lot about these delightful creatures. They, too, are arboreal fruit-eaters but, unlike their red relations, they are small, agile and extremely active. Not for them the wanderer's life of the orang. Gibbons pair for life and live in small, close-knit family groups – a male, his mate and one or two offspring – jealously guarding their compact, fifty-acre territories. Every morning each male directed a barrage of hoots at his neighbours, who took up the challenge and replied with loud calls and spectacular leaping displays. As the excitement built up the females joined in, whooping faster and faster in an incredible trill which I could hardly believe was made by a mammal. Occasionally these confrontations led to fierce battles with animals chasing one another back and forth across invisible treetop boundaries but usually the vocal threats carried sufficient impact to keep trespassers at bay and peace was maintained.

With honour satisfied and their pitch safe from unwelcome

intruders the family moved off in search of ripe figs, the mainstay of their diet. Independent of season, different figs bear fruit at all times of the year. Every day gibbons must range over a large part of their territory checking the trees so that they can reap full benefit from each ripening crop. Arm over arm they swung through the canopy like a troupe of well-schooled acrobats, leaping with careless ease across enormous gaps from one branch to the next. The exuberant young juveniles often led the way with father close behind and the female with her clinging infant bringing up the rear.

Gibbon mothers took great care of their babies, grooming them fastidiously and tickling them in play. From time to time the old male wandered across to see that all was well in the nursery. But it was the juveniles I loved to watch. Performing high among the treetops, swinging and sliding, jumping and climbing, or nonchalantly dangling from the slimmest of twigs, they gave a wonderful display of high speed gymnastics.

As the young animals grow up their father becomes less and less patient with them. Eventually, tired of his aggression, they leave the family to set up on their own, usually loitering on the borders of their parents' range. I often met such lone gibbons on my wanderings in the forest, and provided they kept out of the way of the resident families their presence was tolerated, even when they started calling on their own. It is one thing to overlook a solitary outcast, however, but quite another to accept a member of a rival pair. When one poor male attracted a mate their morning duet brought upon themselves a furious attack from the enraged, local territory holders, who had abandoned their breakfast to hasten to the scene. The youngster was forced to desert his new bride and remain a lonely bachelor until such time as he could find an unoccupied space or fill some vacancy resulting from the death of one of the ruling tyrants.

Although the gibbons were among the commonest

primates in the Bornean jungle, and extremely entertaining, they were very shy and difficult to approach. Whenever they caught sight of me they rushed off to the topmost crowns of the tallest trees and from these impregnable fortresses would abuse me with a deafening riot of derisive hoots until I slunk away from their endless accusation.

The siamangs, or great black gibbons, of Malaya and Sumatra had long aroused my curiosity. Their unusual name and rarity in zoo collections had raised them in my mind to the status of mystery animals. A British zoologist, David Chivers, had just completed a two-year study of siamangs in Pahang and I decided that before I returned to England I must see this unusual ape for myself. Time was short but I was determined to snatch two days to visit the Krau Game Reserve where David had been working.

Unfortunately, my plane from Borneo was delayed, causing me to miss the night sleeper from Singapore to Malaya. I was undeterred and dawn found me on the next train rolling north across the straits. A lift with the Game Department Land-Rover completed the journey and within minutes of my arrival a short, smiling aborigine, Kassim, was hurrying me along a slippery trail into siamang country. The forest here was quite unlike my Sabah area with different tree species and a superabundance of figs and *rambutans*, the like of which I had never seen.

For some two hours we twisted and turned along a maze of narrow paths intersecting the reserve. Leeches were plentiful but seemed small and drab compared to my gaudily-striped Bornean tormentors. We scattered several troops of chattering monkeys, two types of which were quite new to me. The banded leaf-monkey was an elegant fellow with grey back, white waistcoat and black face and hands. Rather more extraordinary was the Chengkong or dusky leaf-monkey, who gives an amazing nasal call rather like the bray of a donkey. With his black face and white clown's rings around the eyes he has a frightening, spooky appearance.

The local gibbons, too, were different from their grey Bornean cousins. They varied in colour from honey to dark chocolate and looked very smart with their white hands and feet and neat white ring framing a black face. I was intrigued to find that gibbons and siamangs, animals of such similar diet and way of life, could live together in harmony, sharing the same ranges and fig trees.

We continued our search and as the day wore on the whine of cicadas rose in volume. From time to time we came upon large Jelutong trees whose trunks bore V-shaped scars inflicted by natives tapping off the sweet white latex from which chewing gum is made.

Far ahead a great tumult began, a tremendous cacophony of booming and screaming.

'Siamang,' called my guide excitedly.

We left the trail and hastened towards the din, wading through clinging swamp and vicious tangles of creepers. After twenty minutes of difficult progress we had still not reached the animals, who sounded so near. We emerged on the bank of a broad, swirling river but, alas, the siamangs were just out of sight on the far side. Kassim ushered me along to the crossing place and we retreated from the loud shrieking and honking that rose in crescendo behind us. Our luck was out. The bridge had been swept away by the flood and when the black apes stopped calling we had no alternative but to return back to the ranger post. I was disappointed and frustrated. It was maddening to have got so close to the animals without seeing them and I had only one more day in which to search but my optimistic companion assured me that I would see siamangs before I left.

Sure enough before the mists had cleared next morning I witnessed a thrilling spectacle. A large, black ape swung with casual indifference into a twisting vine and hung outstretched, accepting my interest as his rightful due.

When he continued on his morning rounds two smaller siamangs followed branch for branch where he led. They were powerful creatures and I had almost to run to keep up. Arms swinging alternately and legs swaying from side to side, they cartwheeled along in an easy, flowing rhythm. This was brachiation perfected; not the slow, careful climbing of the orang-utan nor the risky, arm-propelled leaping of the smaller gibbons but a powerful, confident progression along a perfectly known system of branch connections. Reaching their target they gorged themselves for half an hour on sticky red figs. Then, with one accord, they finished the meal and passed on to an airy *Tualang* tree. There were two adults and two youngsters. The male began to groom the juvenile, thoroughly inspecting his hairy coat, while mother and baby rested nearby, glad to let father take over. The siamang male takes his responsibilities seriously and is more involved in the rearing of the young than the males of other ape species. Second-year infants regularly ride on their father and only return to the female at night. It was some time since this juvenile was so mollycoddled but he still attracted a good deal of fatherly interest.

The group suddenly broke up and the animals began circling the tree excitedly. Like large pink balloons the males' neck pouches inflated and the female's smaller sac was also dilated. A triple bark echoed round the forest like a burst of machine-gun fire and they began in earnest their incredible chorus. Father and son were an incongruous sight, pumping furiously in unison. Deep, gulping booms from their reverberating throat sacs mingled with a string of barks from mother as she led them into another bout of screaming. Several times the calling slowed and I thought it would cease but each time it rose again to a new climax with the frenzied animals leaping among the bare branches like souls possessed. As abruptly as it had begun the chorus ended and the animals sat feeding quietly as though they had played no part in the uproar. Far to the north and west other groups took up the cry, each proclaiming its whereabouts and reiterating its right to be there.

These magnificent apes reminded me very much of the excitable wild chimpanzees I had known in Tanzania. Siamangs are smaller with incredibly long arms but both have the same aura of fierceness and purpose, lacking in the melancholy orang-utan and dainty gibbon. In other respects, of course, they are very different. Chimpanzees live in big, fragmented communities and bands link up and split again as conditions favour. Siamangs, like the other gibbons, live in small family groups, defending their ranges with regular, noisy shouting disputes. Chimpanzee life is full of elaborate greetings and communications necessary to maintain their complex hierarchy but siamangs live so much on top of one another that they have no need of such social niceties. Every member of the family can see exactly what is going on without having to be told.

I watched the entertaining quartet for two enjoyable hours before rushing back to the Game Post, where the Land-Rover was waiting to whisk me to the station. Soon I was speeding south again and a day later I was flying back to the cold shock of another English winter.

I returned to Oxford, where life was very busy. I had to write up the findings of my Bornean trip and spent a hectic time giving seminars and film shows to academic and conservation-minded audiences. Moreover, I was trying to obtain funds and permission for another trip to the Far East. My brief glimpse of the siamangs in Malaya had aroused in me a gnawing desire to visit the only region where the ranges of siamangs and orang-utans overlap. In North Sumatra both occur and share their forest habitat with the gibbons. I was anxious to see how three fruit-eating apes could coexist. Did they cohabit peacefully or were there terrible inter-species jealousies? Did the Sumatran orang modify his ways in order to live with the siamang or was he just the same as his Bornean cousins? Such questions could only be answered in the field.

Because of the many different Indonesian authorities who had to be consulted and the great postal strike in Britain it took me several weary months to obtain permission to go ahead with my plans. The delay was useful, however, for it gave me time to go into hospital for a thorough check-up. I had never fully recovered from my winter in the jungle and now had to pay the additional price of donating what seemed like gallons of blood to pathology laboratories all over the country. The list of extraordinary illnesses from which I was suffering grew and grew until it was quite a joke among the hospital staff. As each disease was diagnosed it was treated, but I still felt ill and now had to bear the ignominy of becoming number one exhibit to the medical students. A pint of beer was offered to anyone who could come up with the answer but the drink remained unclaimed for the medical officer himself decided to try a new test and discovered that I was harbouring a parasite normally found only in dogs! Even when this had been cured a raging dose of malaria forced me back into a hospital bed two weeks later. I was obviously in no condition to defend myself against the pretty blonde who came to visit me every day. Kathy and I got engaged a few days before I flew to Singapore with promises that she would join me in the autumn and we would marry in Indonesia.

Part Three

Sumatra 1971

Chapter nine

Sumatran Reconnaissance

It was not until the beginning of May, after further confusion over my work permits, that I at last saw Medan, the capital of North Sumatra. Medan is a big, sprawling town with a population ranging from half to two million, depending on where you draw its boundaries. It was an enormous change from Borneo for Sabah's entire population numbered only 600,000. The principal means of transport was the *betcha*, a sort of bicycle carriage derived from the Chinese rickshaw. The long Dutch influence in North Sumatra was reflected by horse-drawn traps, western vegetables and curious sheep and cows that cluttered the back streets.

Like the rest of Indonesia, Medan had two levels of economy. Hotel accommodation or food in a restaurant was very expensive but a room in a small *losman* or a meal from a roadside stall cost virtually nothing. I was spared the inconvenience of either style by an invitation to stay with the family of a Wildlife Conservation Officer on the outskirts of town. Thus began my introduction to the Indonesian way of life. The food was so hot with chilli peppers that I was quite unable to speak at mealtimes. This was probably just as well for my Bornean Malay was considered rather quaint

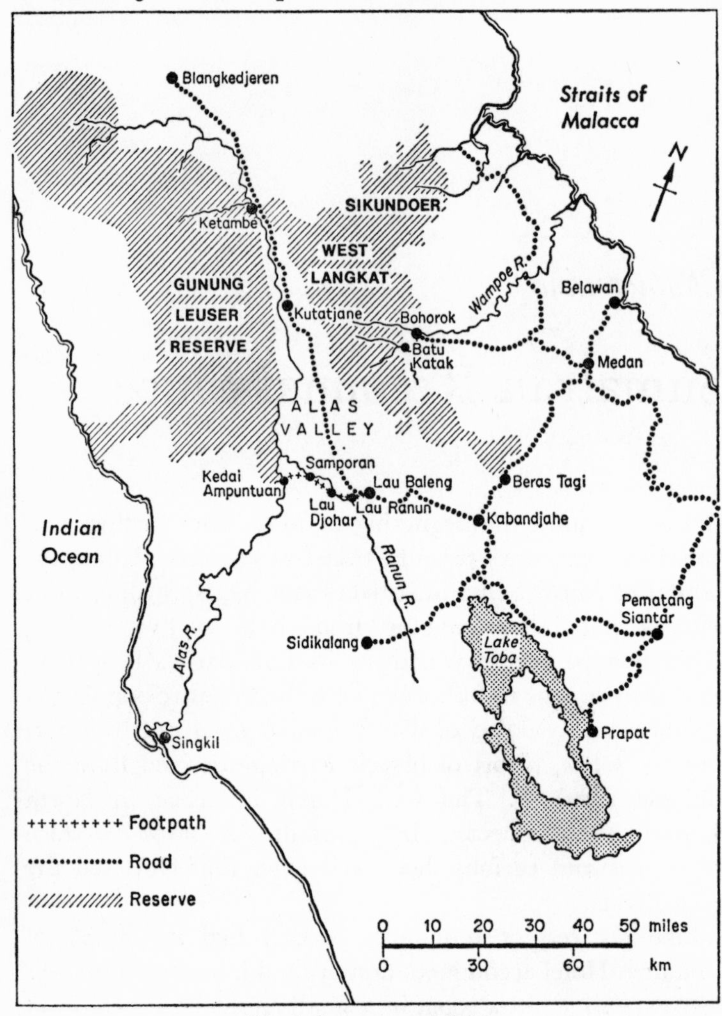

North Sumatra and Atjeh

in Sumatra and people roared with laughter whenever I opened my mouth.

At the back of the house was the typical Indonesian bathroom or *mandi*. This was a small room, some eight feet by twelve, housing in one corner a tall, concrete water tub provided with a tin scoop for pouring water over oneself.

The spillage drained away down a large hole in the centre of the floor which also served as a latrine. In Mr Manalu's house the arrangement was quite standard but with one exception. The *mandi* was occupied by the stout household pig. I was assured that she was a friendly and good-natured creature but she had obviously never seen a white skin before. On my first visit the pig charged round the small space, emitting deafening squeals, and nipped me sharply on the leg as it passed. I hastily completed my ablutions but future calls of nature became a frightening embarrassment. My mind retains a vivid picture of my naked form squatting in the middle of the floor while I expected to be attacked at any moment by the fat beast who eyed me meanly from the far corner and snorted her disapproval.

My visas and permits were still not ready and I had become so frustrated by life in town that without further ado I collected my kit ready for a trip into the forest. I had to find a suitable study area for my work and wanted to survey various places. I had heard favourable reports of the huge Gunung Leuser Reserve in Atjeh province. Nearer to Medan, however, was the West Langkat Reserve and I decided to make a short trip through this region to see if it was at all promising. The Conservation Department lent me a ranger and arranged a lift for us as far as the township of Bohorok, close to the mountain reserve. In Bohorok I collected two more men and the everyday essentials like salt, chillis, dried fish and rice.

Early next morning we set off on foot through the rubber plantations and neat, little *kampongs*, shaded by tall kanary nut trees. By ten o'clock we had reached the boundary and were following the course of the Landak (Porcupine) River. On either side the forest-clad hills rose steeply and were alive with the furious chorusing of the gibbons and siamangs. We branched off along a small stream which tumbled down from the ridge in a series of tiny waterfalls. Great hornbills sailed overhead in the clear sky and from an open slope we could gaze out over endless miles of leafy canopy. It was wonderful to be back in the jungle again after so many months in civilisation. The treetops shook at the passing of a family of gibbons, their every movement punctuated with sudden flashes of white, for these were white-handed gibbons similar to those I had met briefly in Malaya.

We crossed the ridge and began the steep descent into the adjoining dale, following deep trails made by water buffalo as they dragged logs cut by timber poachers. Such illegal felling is a major problem in the Sumatran reserves but their boundaries are so extensive and the conservation departments so short-staffed that it is difficult to protect them. Finding a shallow stretch, we forded the fast-flowing Berkail River and walked cautiously along its slippery bank. The thick forest was dark and damp and the ground so steep that our progress was painfully slow. Roaring and splashing over the boulders, the mountain torrent made such a din that we could hear little else. As this was absolutely useless for finding animals, I decided that we would do better to follow the ridges so we headed wearily uphill once more. Up and up rose the slope, so sharply that at times we had to crawl on hands and knees or haul ourselves forward by clinging on to bushes and vines. With a short halt for a much-needed rest and a meal we scrambled on up the sheer face.

It was already evening when we reached the top of the ridge. The ascent of some two thousand feet had taken four hours. The hard climb had been thirsty work but we had little water left so, leaving the men to construct a shelter, I wandered off in search of further supplies. My quest proved difficult for any rain soon drained away on such a sharp gradient and all the stream beds were dry. I had no alternative but to descend the trail again for a considerable distance before I found a shallow puddle and could fill the bottles. The forest was now blanketed in darkness and an ominous clap of thunder heralded the start of a torrential downpour.

Before I could reach the friendly cover of our camp I was soaked to the skin by the sudden flood. Even more maddening, now that the rain had come, my trip had been in vain. A steady stream of fragrant, fresh water poured off the polythene roof into a strategically placed cooking pot and I could throw away my muddy offering.

Lightning flashed and the trees swayed wildly in a wind that had sprung up from nowhere. With a loud crack a heavy branch tore free and hurtled to the ground. Suddenly we froze in silence as an incredible roaring began within a hundred yards of us. Deep and terrifying, the bellows continued for a good half-minute or more. The men were scared, assuring me that this was the voice of *Harimau*, the tiger. It certainly sounded like some enraged large cat but I was struck by its amazing similarity to the call of the Bornean male orang-utan. Couldn't it be a *Mawas*? I asked, but they were adamant. This was *Harimau*.

Ignoring the rain we lit a fire and cooked our ration of rice. The embers glowed comfortingly through the night and, exhausted by our exertions, we slept soundly while the tempest raged around us. It was still dark when I awoke but the rain had ceased and dawn could not be far away. Hombing, the game ranger, was already up and puffing a cigarette as he gazed out at the dripping forest. Without warning the roaring started up again and this time I was sure it must be an orang. No tiger would have stayed in the same spot all night to sit out such a storm. Although the call was much faster and shorter than that of the Bornean male it was so similar in other respects that the unseen challenger had to be a *Mawas*.

We crept through the gloomy jungle towards the sound but all was now quiet. Two fresh nests caught my eye. There was no doubt they had been used for the night but they now stood empty. It was the small juvenile we saw first, feeding quietly only a few yards from us. Hearing our approach he retreated into some taller trees and from this sanctuary squealed at us and kissed his wrist, a most unusual display. Branches moved ahead, betraying an old male who was swinging a low bush to and fro until it gained sufficient momentum to carry him across the gap to the next support. Close by but higher sat a plump female. The two adults climbed into an open tree to feast on the epiphytic ferns which festooned the boughs. Gradually the male edged closer to his partner and suddenly lunged forward and grabbed her. Squeaking with mingled fear and pleasure, she allowed him to mount her and for nearly ten minutes they continued their love-making unashamedly in front of us.

They broke off to feed once more but, seeing us peeping Toms, the ardent lover bellowed furiously and began shaking branches in protest at our rudeness. His mate took this opportunity to slip away but her departure did not pass unnoticed and the male was soon hurrying after. The juvenile made a wide detour behind us and rejoined his elders, whose squeals and grunts proclaimed another conjugal bout. We followed after the happy family and caught a few more glimpses of their private habits but the ground sloped steeply away and eventually we had to turn wearily back to camp, where our companions still slumbered. It was wonderful luck that my very first Sumatran orangs should be indulging in a mating session for it had taken several months before I had witnessed such sights in Borneo. I had been somewhat anxious about the wisdom of changing islands, rather than returning to Sabah for a reappraisal of the situation there, but now everything seemed to bode well for the Sumatran study.

For six more days we travelled in a wide loop through the mountains. When it became quite impossible to get water on the high ridges, we were forced back down into the valleys, taking a whole day to descend only a mile on the map, so terribly steep was the terrain. I had to remove my shoes to scramble over the slippery boulders of treacherous waterfalls and my feet were soon cut and sore.

We found three more groups of orang-utans and enjoyed many other exciting glimpses of the local wild-life. In a pouring rainstorm one huge orang-utan came swinging menacingly towards me. I tried to call his bluff by running at him, shaking my cape from side to side. But far from checking his advance this action seemed to spur on his attack and the great animal hurried to the ground to meet me. I turned tail and ran for all I was worth, fell, picked myself up and dashed on again.

Fish from the tumbling rivulets supplemented our fast-declining food reserves and by the end of the trip we were all very fit from our severe exercise. The area was the steepest and roughest stretch of country I had ever attempted and our packs had been far from light. We finally emerged from the reserve and, aided by stout stakes, waded slowly through the tugging currents of the Berkail River. On the other side a path twisted and turned through a fairyland of limestone caves and grottos. Flimsy ladders traversed yawning gullies and the whole place had the unreal appearance of a scene from Aladdin. The path opened on to a small clearing, where stood the twelve stilted houses that comprised the *kampong* of Batu Katak (Frog Rock). We celebrated our return from the forest by purchasing a chicken and devouring it speedily, then retired to a vacant hut for the night.

In the small hours we were roused by a great commotion. Lamps bobbed and vanished as the villagers dashed to the scene of the excitement. An enormous python had slid into one of the houses and the owner and his friends were busy trying to catch it. Since they were anxious to sell the fine skin they left the snake unharmed and persuaded it into an old flour sack. I inquired how much it was worth, added a small percentage and was duly presented with 'Si-Ross' (named after the flour company), although I had no idea what to do with my latest pet.

Armed with lanterns and torches, we returned next morning to investigate the incredible limestone formation we had passed the day before. Once there had been a regular path up the valley but this had fallen into disuse and the dank jungle had encroached to such an extent that we found it easier to splash up the stream bed. Even then we often had to resort to our *parangs* to hack a passage through the dense vegetation. At last we reached a small waterfall and just above entered the lofty caverns and tunnels that honeycombed the mountain. The domes of the vast cathedrals twittered to the harsh squeaks of thousands of roosting bats and their tiny eyes glowed red in the torchlight. Clicking swiftlets circled and swooped around their saliva nests, high on the rocky walls. We crunched over deep piles of slimy guano, embedded in which were the decaying skeletons of dead bats and fallen fledglings. Every few yards we made sure our trail was well marked for it would have been easy to lose direction in this maze. At times we had to crawl on our stomachs along low corridors or wade, thigh deep, through pools of icy water. We emerged into huge chambers, where stalactites and stalagmites reached to each other for cold comfort and the dripping rock icicles stretched upwards to fathomless heights.

Excavations in limestone caves in Central Sumatra have yielded thousands of teeth from prehistoric orangs who probably came there to shelter. I hoped that these caverns might conceal similar zoological treasures and we wandered round the pitted cliff face diligently searching at the mouths of several black holes. Although we disturbed a few sleepy porcupines and basking lizards, our archaeological discoveries were limited to one blackened layer of soil, which suggested the use of fires in times long past. In spite of our lack of success I am sure that the region merits further more serious investigations.

Back at Batu Katak my snake and sack had disappeared. We searched everywhere but it was not until I glanced despairingly upwards that I spotted the bag twisted round a rafter. In spite of his confining encumbrance the python had managed to wriggle up the roof pole and if we had been away much longer would no doubt have made good his escape. Before he could try any more tricks we gathered our belongings and climbed over the hills to Bohorok and thence to Medan.

I was happy with the results of the Langkat trip. We had seen ten orang-utans and a wealth of other animals. The harsh ruggedness of the wild country had been greatly stimulating and I now felt fit enough for anything. It was clear though that this was not the area to carry out a more detailed study. Travel was difficult and the roaring rivers and waterfalls obscured the delicate sounds on which any naturalist must rely. It seemed that I would have to visit the Gunung Leuser Reserve farther west, beyond the broad Alas River.

The journey to Kutatjane was an experience I should not care to repeat often. For ten long hours I was jostled and shaken in an ancient, gaudily painted bus, travelling dangerously fast along a road scarred with terrible potholes. I should never have believed that so many people could cram aboard one vehicle and I spent an uncomfortable time trying to find room for my knees. Even the aisle was bridged by boards so that not an inch of valuable seating space was lost. The conductor had a most hazardous job, for to collect the fares he had to inch precariously along a rail outside the bus, passing change through the glassless windows. Yet more passengers were perched on the roof amid all the luggage, protesting livestock and bulging fruit baskets.

We rumbled up the steep, tree-clad escarpment of Sibolangit, passed the great volcanoes by Beras Tagi and Kabandjahe and crossed the arid *blang* (waste) country to the north of Lake Toba. At length we descended into the Alas valley and hurtled onwards to the northern province of Atjeh. The wide flood plain was completely cultivated and the women were busy hoeing and planting in the *padi* fields. On either side the forested mountains of the Langkat and Gunung Leuser Reserves reached skywards. Our transport clattered through the Karo *kampongs*, where bare-breasted women and fat domestic pigs attested to the fact that the Muslim faith had not taken hold in this part of Sumatra. Without stopping, the driver yelled instructions and hurled letters to anyone standing by the roadside as we dashed past. This is the only way that post can be delivered in these areas and it is surprisingly efficient.

Towards evening we pulled up in the main street of Kutatjane and I found a room in the Government Hospital Mess. I had to present myself to the various officials of the Police, Army and Forestry Department to pay my respects and explain my presence in Atjeh. After opening an account in the local bank and emerging with a thick wedge of transaction books that belied my modest investment, I set about finding some way into the forest. A rough road runs the sixty miles from Kutatjane north to the town of Blangkedjeren but there is no bus service. The track follows the course of the Alas River, which serves as the boundary of the Gunung Leuser Reserve. While I was keen to enter the reserve, I found I would need both a jeep to manage the road and a boat to cross the fast-flowing Alas. Neither was readily available and funds for such luxuries were short.

To the south of Kutatjane, however, was a large block of forest on the near side of the river and within walking distance of the main highway. Although this region was not within the confines of the reserve, it was still unspoiled and known to harbour orang-utans and rhinoceros. A long-used path provided easy access from the village of Lau Baleng. As there are no roads across North Sumatra the only way that goods can reach Singkil on the west coast is either by a laborious boat trip round the northern tip of the island or, more commonly, by porters or *burohs* following the trail from Lau Baleng to Kedai Ampuntuan, a boat station on the Alas. From here the goods can be floated down-river to Singkil. For the princely sum of a thousand rupiahs (£1) a *buroh* will carry a load of over a hundred pounds (50 kg.) the twenty miles to Kedai Ampuntuan.

I braved the local bus once more and hastened to Lau Baleng to make myself known to Mr Pardede, the cheerful head forester. His assistant, Taragan, volunteered to accompany me on a brief survey, but not without the department rifle. Encumbered with this awkward and dangerous-looking veteran, we staggered off with our heavy loads along the muddy buffalo track to the Ranun River. We crossed the hot, open, sun-drenched hills to the village of Lau Ranun and passed through shaded banana plantations to meet the river again at Lau Djohar. Here we found the elder Nami, who at well over sixty was still fit enough for the jungle and knew these parts better than any man. We persuaded him to join our expedition and had an enjoyable meal with his family in their palm-thatched house.

Nami told us that only ten days before a tiger had killed a cow. The local policeman, determined to destroy the big cat, had waited all night in a tree-hide overlooking the kill. The tiger had returned for another meal but the policeman was so excited at this chance that he pulled the trigger long before he had sighted on his potential trophy. A cow elephant and baby had been seen in the nearby Djohar valley and would probably still be in the vicinity. We decided to investigate this report at once.

The Djohar stream ran down to the Ranun along a broad valley watched over by the mountain of Gunung Umang at its head. We followed the water-course, stepping lightly over the green and white flecked boulders that littered the shallow bed. The crystal clear water tinkled happily down a thousand tiny waterfalls to create an enchanted dell. Chains of saffron butterflies danced over the pools and a blue-banded kingfisher skimmed ahead of us to watch with bobbing bill from a safer distance. We soon encountered the tracks of the elephant and her tiny calf but they were several days old and we continued in search of something more recent.

By the time we reached the top of the valley the sky was darkening. The Katydids struck up their mechanical whining and we hurriedly looked about for a campsite. Taragan and I cut saplings to construct a rough shelter while Nami climbed down to the stream for water. We chewed our toasted, salty fish with raw onions and soggy rice, washed down with meagre mugs of bitter coffee, and settled for the night. At 2.30 a.m. I heard the faint sighs of a distant orang and took a compass bearing. With daybreak we set off on the magical line, which I assured my companions would lead us to a large, male orang. Old Nami shook his head doubtfully at such new-fangled notions. Twenty minutes later we were standing on the brink of a steep precipice, my compass needle pointing confidently over the edge.

It took us a long time to scramble down to the foot of the cliff still heading roughly in the right direction, so it was rather frustrating to find that the sheer face curved round to block our path again and we had to reascend the crumbling rock if we wished to maintain our course. I finally decided to abandon the chase when the needle pointed an impossible route across a perpendicular drop and we cut our losses by wandering straight up the ridge to the slopes of Gunung Umang. The mountain derives its name from the mythical little folk who supposedly haunt its summit, but the only sign of men we found were the *parang* slashes along the paths made by collectors of wild cinnamon. Here Nami took over the navigation and soon had us back on familiar terrain.

We marched west along the trail to the Alas River. Small clearings had been cut every two miles or so as resting places for the weary porters. I travelled first to catch sight of any animals we might disturb before they could flee or hide. We passed three orang-utan nests but they were brown with age. Grey leaf-monkeys cackled with alarm and fled away through the swaying branches and red-crested pheasants scuttled among the dry leaves. It was a delightful area, rather lighter and drier than other forests I had known. The path ran parallel to the river, sometimes meandering along the bank, sometimes overhanging it on the edge of the steep scarp.

I stepped out to cross a small stream and shrieked with surprise. Peals of laughter came from the men behind. The water was as warm as I could bear and a short way up its course I found a hot spring bubbling out of a crevice in the rocks. Nami explained that the temperature of the rivulet varied according to the weather, curiously being hottest when there had been a lot of rain. He said that the waters were excellent medicine for stomach upsets.

We continued on our weary way, up and down, down and up, following the trampled paths of deer and elephant across the many streams that tumbled to meet the Ranun River on our right. Suddenly, in the middle of miles of jungle, we came into a small *ladang*, planted with tapioca, pawpaws, chilli peppers and aubergines. A wooden *pondok* (hut) stood in the middle of the clearing and fresh ashes and fish bones indicated that someone lived here. Nami declared that the house was a night stop on the long walk to Kedai Ampuntuan. It had been built by a friend of his two years earlier and the owner had stayed there, making a living by selling food and coffee to the professional porters who came through. Since the police post had been set up at Lau Djohar, however, the *burohs* either had to give heavy bribes or pay for an expensive permit to travel between the provinces of Atjeh and Sumatra Utara and the numbers making the trip had declined considerably. The coffee house was no longer profitable and the proprietor had returned to the *kampong*. When I pointed to the fire Nami laughed. 'Oh, that is old Turut, tuan. He is a hermit who sometimes stays here, catching fish in the river or seeking wild cinnamon in the hills.' Nami said the place was called

'Samporan' (waterfall) and pointed to a track winding back almost the way we had come. I could, indeed, hear a waterfall in the distance. We decided to spend the night at the *pondok* and I wandered off to inspect the falls. It was farther than I expected but a thunderous roar proclaimed that I was drawing close. The path descended a steep gully through a twisting corridor in the rocky outcrop and wound along the bottom of the cliff to a sandy beach, sheltered by a clump of giant bamboos.

The river swirled in a great brown pool but I had to clamber over the cliff face before I could see it to full advantage. The torrent was channelled through a narrow cutting and splashed in three sudden steps, changing course and plunging down a fifty foot drop into a foaming, boiling spume. Strange ferns hung from the ledges, their leaves moistened by the cloud of rising spray. Delicate pink orchids peeped from among the rubble of lichen-covered boulders and on the beach itself was the most impressive sight of all, four Rajah Brooke butterflies drinking on the wet sand. They flickered in lazy circles and landed gently to suck up the sweet water through their long proboscises. Metal green shimmered on their six inches of black, velvet wings and, with their scarlet collars and flashes of blue beneath, they were truly the handsomest butterflies I had ever seen. One spiralled above me and sailed away over the treetops but returned a few minutes later in a perfectly controlled, gliding dive that was simply breathtaking.

Delighted by my find I hurried back towards the *ladang* but my thoughts were soon diverted by the barks and screams of a family of siamangs warming up for a chorus. I left the track and crept quietly towards them. I could see two adults, and a juvenile, all in one tree, which shook with the vigour of their calls. In and out puffed the strange, grey throat pouch as the male gave his characteristic gulping boom. The female began her series of loud barks, which so excited her partner that he leapt frantically round the tree, accompanying her with high-pitched screams. The juvenile, too, seemed very agitated but did not join in the deafening crescendo. For several minutes they continued their urgent song until the juvenile spotted me and scurried off. Alerted, the rest of the family forsook their wonderful calling and hastily followed suit, their long black arms swinging like cartwheels.

From Samporan to the Alas River the path was even more tortuous, doubling back on itself to circle round craggy peaks and swampy areas. We found a cluster of orang nests congregated round a strangling fig but met only gibbons and two herds of wild pigs. After a long uphill climb, and an even sharper descent, we reached Kedai Ampuntuan, marked by a solitary hut beside the river. Once this had been a bustling station with numerous boats calling every day to collect the portered goods and ship them down to Singkil. Now, however, only one man remained there and craft were rare. A party of travellers had already waited in vain for two days and I had to give up hope of being ferried across the Alas.

It was durian season at Kedai Ampuntuan. Eight fine trees shaded the hut, at a safe distance from the encroaching forest and the evil designs of roaming orang-utans. Every few minutes a loud thud announced the fall of another ripe fruit and one of the group would rush out to collect it, praying that another did not land on his head. Five pounds of spiny durian dropped from a hundred feet is a formidable missile and nasty accidents sometimes occur. Back in the safety of the house we prised apart the prickly shells and shared the ripe contents. Chestnutlike stones were covered in sweet, yellow aril, whose sickly smell, smooth, creamy texture and indescribable combination of flavours have made this to some the most desirable, and to others the most loathsome, fruit on earth.

There seemed little hope of finding a boat to take me across the Alas so I decided to return to the 'village' of orang-utan nests we had passed along the way. Early in the morning we staggered back up the steep hill behind the *kedai* and headed back to the twinkling stream of Liang Djering. The men set up camp here whilst I went off into the hills in search of our elusive quarry.

The area was traversed by wide elephant trails and in a mud wallow I found deep prints where the lordly beasts had bathed a few days earlier. Luck was with me again for I soon found orang-utans and during the next four days was able to spend many hours watching six new subjects. Most entertaining were a charming trio, a young red male accompanied by a plump female, who cradled a tiny golden baby. In the evening sunlight the family made a gorgeous scene, their russet hair glinting amid the dark green vegetation as they munched their way through a feast of scarlet *sasaga* beans that hung in curly pods.

It was time to return to Lau Baleng and make plans for a longer stay. I was favourably impressed with the Ranun region and had decided to make this my study area. Travel was easy along a useful system of natural paths. Orang-utans and other primates were plentiful and the deserted *ladang* proved an ideal camping site which would provide us with fresh vegetables and pawpaws. A calling family of siamangs lived right on our doorstep and I was greatly intrigued by the magnificent waterfall, hot springs, towering white cliffs and Rajah Brooke butterflies. I could hardly wait to get my luggage brought down and a new house constructed.

Chapter ten

Pythons and Elephant Caves

I decided to keep on old Nami and befriended the local policeman, a likeable villain, by hiring his cousin, Purba. Purba was a short, stocky fellow, powerfully muscled from his years of working in a timber camp. He also had the dubious qualification of having been cook to a band of guerillas who had roamed the jungle for three years. Nevertheless, he was more of a town and *kampong* man than a forest dweller and had little detailed knowledge of the animals and plants with which our world abounded.

A buffalo cart transported my heavy trunk, Si-Ross, the python, our tins of rice and other supplies from the Pardedes' house to the village of Lau Ranun. Here I paid for a boat to carry the equipment as far downstream as possible. Because of the waterfalls everything had to be carried the last mile to camp. Purba and Turut, the hermit, cut a path from the landing site while Nami and I chopped poles and bound them together with strong liana fibre to build the superstructure of our hut. Also from Lau Ranun came sixty lengths of grass thatch that I had ordered. Strips of bamboo tied to a skirt of grass were fitted like tiles to form a firm, waterproof roof. Turut built a springy sleeping platform of split bamboo and, with the roofing fixed firmly in place with

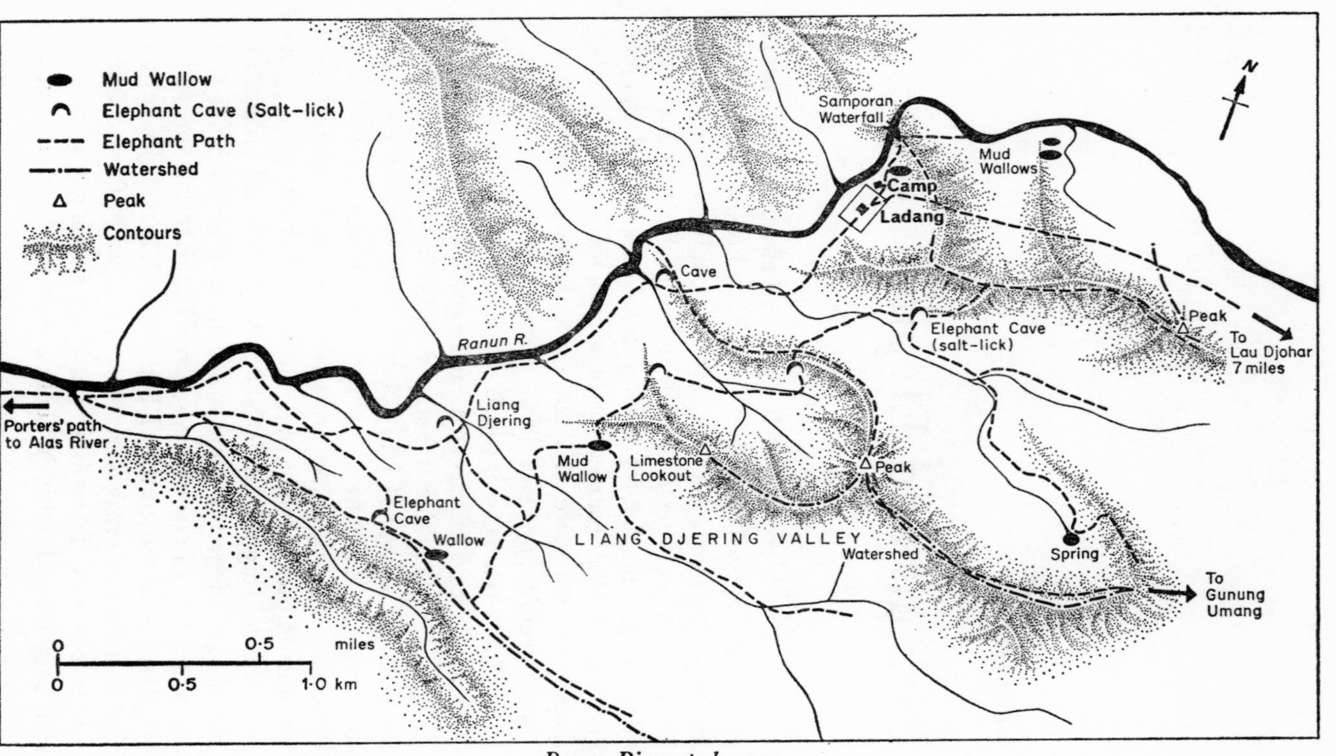

Ranun River study area

rotan cane, our little home was ready. Purba made a table and bench in front of the house and here I dined in kingly fashion. With the work complete Turut wandered off on another of his forays but Nami and Purba stayed to supervise the camp affairs, leaving me free for my work in the forest.

One of the most apparent features of the Ranun area was its active population of elephants. A network of broad, clear paths and piles of dung served as a reminder of their passing. Beautiful fungi, silvery blue caps supported on slender stalks, adorned the droppings, helping them to decay and giving some clue as to their age. Trampled *Caryota* palms, crushed bamboo and torn lianas showed where the great beasts had fed and mud smeared high on tree trunks marked their rubbing posts. Most interesting of all were the remarkable caves that the elephants had gouged in the hillsides to obtain the white pumice stone, rich in salt. I knew of five different sets of caves and all were regularly visited. Many were as much as twelve feet high and twenty-five feet deep. Their roofs were pocked with long, smooth tusk marks where the bulls had stabbed again and again to break off fresh rock. Lacking tusks, the cows and calves had to be content kicking the soft strata till they crumbled. Elephants ate huge quantities of this pasty soil and for a couple of days afterwards their dung would consist mainly of sand.

Other animals took advantage of these excavations and left their teethmarks on the walls and footprints. On the dusty floor, I found the hooked casts of visiting orang-utans and the strange, brush trail of the porcupine, slots of barking and sambur deer and prints of leaf-monkeys and squirrels. Even the treetop gibbons came to eat there. Their teethmarks were high under the entrance of the largest cave and a low branch was worn smooth by the hundreds of hands that had gripped it while siamangs and gibbons collected the rare minerals.

At the main set of caverns I built a ladder and tree-platform from which to spy on the activity below. When a group of six elephants toured the area I spent one and a half days up that tree. Each time I hurried to the caves I seemed to disturb the giants feasting there and they hurried away. Throughout the day I could hear them crashing about in the surrounding undergrowth but none dared approach while the strange smell of man lingered on the air. I saw a deer and a herd of wild pigs but was unable to photograph the kings of the jungle as I had hoped.

In other parts of the forest the elephants were no tamer and always moved away when they caught my scent. Eight elephants used the area and ranged as far as Lau Djohar to the east and the heights of Gunung Umang in the south. About every six weeks the animals passed through my territory, staying a week or so before moving on in their clockwise perambulations. Two cows and their calves often travelled together and once, when I met them in the hills, they had a sub-adult male with them, but the other three elephants usually ranged alone. One of these, a young bull, entered our *ladang* one evening and moved up slowly behind camp, browsing on the young bushes that had sprung up since the clearing was cut. Another very big male left his tracks and dung near the hot springs outside my boundary but I never saw him.

On a trip up Gunung Umang I spotted what I at first took to be an orang-utan. For several minutes I sat watching a low tree which swayed as some large animal fed among its branches. I could see a long, hairy arm picking off leaves but when the animal emerged from behind the screen of bushes I realised that it was the trunk of a baby elephant that I had been observing and he was closely followed by his mother, a formidable giant. As mothers with such small calves can be quite fierce I stole quietly away and left them undisturbed.

The belief that I had magical powers over the forest animals was common among the superstitious *kampong* folk but I was rather taken aback when an official from the

Forestry Department arrived to ask me to call away the elephants that were damaging the rice fields near Lau Djohar. I tried to explain that I was as helpless as any other man and recounted the occasion when Purba and I had crossed the river to investigate orang calls which had disturbed us during the night.

Nami rowed us across in my little boat. Purba, who could not swim and had never been in such a small *sampan* before, was stiff with fright and almost swamped us with his jerky movements. Safely deposited on the far shore, we headed along a clear trail into the hills. Although there were several orang nests none was new and we circled back towards the landing site. Loud cracks of splintering bamboo warned us that an elephant was feeding nearby. As Purba had never seen an elephant we crept slowly towards the sound. From only a few yards away we still could not see the animal so I indicated a small tree with a convenient hanging liana and we started to climb. Suddenly the bamboo curtain burst apart like an exploding matchbox as a huge-tusked bull thundered down on us. We were up the tree in a flash. Purba whimpered pathetically as we clung to the flimsy, sagging branches, while the monster stamped beneath, eyeing us meanly and waving his long trunk in our direction. With an irate bellow he stormed off through the forest, trumpeting as he went. We tumbled to the ground like a couple of jellies and hurried back to the river as fast as our shaky legs would carry us. Purba was only too keen to clamber into the little boat that had so terrified him earlier. Over the next few weeks he made a good deal of the story but he was never tempted to cross the river again.

Tigers lived in the area but they were so few that they posed little real threat to the forest worker. Even so, the very fact of their existence made a lot of difference to my sense of ease. In Sumatra I never slept out alone in the forest and even with company always kept a fire burning through the night, a precaution I should not have dreamed of in Borneo. One day I heard a tiger roaring in the broad valley at the western end of my range; it was an eerie noise, not unlike someone trying to start an obstinate motor-bike. When I explored I was assailed by the powerful odour of frightened snakes and Turut later found the half-eaten body of a cobra. Within the week I discovered bones of a recently killed pig but we never saw the culprit. Perhaps it was the same tiger that had attacked the cow at Lau Djohar but he seemed to stay in our region for only a few days, after which we came across no more signs.

Smaller cats were not uncommon and I sometimes found their prints on sandy river banks. One night we heard an animal scrabbling behind the house and I shone my torch into the inky shadows. Two bright eyes blazed at me before the creature turned and fled. I could not tell whether it was a panther or the slighter golden cat but I had no doubt that it was the smell of our piglets and chickens that had attracted his unwelcome attention.

On my ramblings I came across an open window where a landslip had cleared a gap in the screening vegetation. About a mile away a tall, white cliff rose sheer out of the clinging forest, its treeless peak silhouetted stark and unfriendly against a stormy sky. Intrigued by this impressive formation, I scrambled down the slope, splashed through the gurgling stream and headed towards the wall of rock, guided by my invaluable compass. It was a steep climb strewn with huge, jagged boulders of hard limestone but I pulled myself up on the thorny bushes that somehow managed to retain their hold on the sloping terrain. I was already several hundred feet above the river when I emerged from the prickly scrub at the bottom of the cliff. A narrow crack ran diagonally across the face, providing a haven for clinging ferns and orchids, and as the rock seemed firm I decided to give it a try.

The limestone had eroded in curious pits bordered by sharp, knife-edged blades of stone that threatened to cut through the thin soles of my shoes. A single slip, a short fall and I would have been cut to pieces. I was terrified by this prospect but since the way back looked even more treacherous than the road ahead I was committed to go on. I inched cautiously along the ledge and, when it petered out to nothingness, clambered on to the next perilous shelf. With enormous relief I finally crawled on to the summit of my ridiculous conquest. There was nowhere to sit on the sharp, pocked crag but the high vantage point offered a wonderful panorama of the whole region. The valley of Liang Djering slept below; to the west I could see the hills beyond the Alas River, to the north the sweeping mountains of the Langkat and Leuser reserves and rising behind me the rounded peak of Gunung Umang.

Lichens and a handful of pretty orchids were the only plants able to gain a foothold on the bare pavement. I marvelled at their tenacity. Their roots had insinuated themselves into the shallowest of crevices and, as they swelled, widened the gaps to make room for further colonisations. I was gratified to find an easier path descended the far side so that future visits to the Limestone Lookout need not be so strenuous. In a worn hollow, old droppings told where mountain goats were wont to shelter. I slithered down among the jutting slabs, sustaining my only scratch of the whole escapade, and wandered back into the cool forest.

By now camp life had developed a smooth routine. Purba cooked and cleaned clothes, while Nami lovingly tended the *ladang*, planting maize and onions or picking tapioca leaves for our stews. Nami had arranged for a few chickens and two piglets to join us so that we should never be short of food. The maize grew taller but as it did so it attracted the wild pigs from the jungle. Beneath its protective cover they dug up the tapioca roots and, despite all Nami's valiant efforts, had soon finished the crop. Nami spent much time in building some ingenious traps but the wary beasts always eluded them and we never got the better of them.

Another *ladang* raider was a male pig-tailed macaque. This large monkey was as big as a baboon with a frightening expression and long, sharp teeth. He was passionately fond of young maize cobs and daily left his troop to creep unnoticed among the tall corn. Slowly he worked his way through the field, biting out the juicy cobs, one by one, as they started to swell. Our only defence was to pick them unripe for the monkey was much too clever to be dissuaded from his trick and was rarely spotted although his forays were always during the daytime. Often he perched in a tall tree at the edge of the forest, waiting for our attention to be distracted so that he might make another quick sortie.

Turut took up residence in the old house in the *ladang*. He was a wizened little man, a catcher of fish and a trapper of birds. Most of the time he caught scarcely enough to live on but occasionally the river rose, bringing up shoals of new fish from downstream, and Turut would dry his catch and exchange it for a few months' supply of rice and salt. Loud and happy, he sang away the night in his lonely hut but daybreak found him scampering over the rocks, baiting his lines with fat, green crickets or succulent grubs from rotten wood. Turut had an amazing knowledge of the forest plants. He knew more species of edible leaves and fruit than anyone I had ever met and could always be relied upon to produce some greenery to liven up a monotonous diet.

When fishing was poor he would turn his attention to birds. In Sumatra the familiar office rubber plant grows wild as one of the biggest trees in the jungle. From its sticky sap Turut made bird lime for catching small birds. Larger game, such as pheasants and partridges, he trapped with fishing-gut foot-snares. He carefully cut saplings and bushes to make a long, low wall with gaps every few yards, where cleverly placed twigs guided the ground-walking birds over the loops. As they trod on the *rotan* mats buried beneath the leaves they released the trigger of a bent twig, which sprung back, pulling the noose tight round the foot of the victim.

By buying fish from Turut at about twice the normal town price I was able to distract him from his bird activities but even so he found time to catch a few partridges.

Wednesday was market day in Lau Baleng and every week Purba made the long trip to buy supplies and collect my mail. Usually he purchased far more than he could carry himself and had to hire men to help transport the pile of rice and vegetables back to Samporan. One evening he arrived in camp, loaded with goods and accompanied by a friend similarly burdened. It was already after dark and as they had no torch they had run the last mile to reach home while they could still see the trail. Hot and panting, Purba told me that the third member of their party was still in the forest. Since he had a club-foot their companion could not keep up with them but he was not afraid to travel alone and should soon arrive. Like them he had no light and not even a *parang* for comfort.

It was foolish of them to have left so late. They should have spent the night in Lau Djohar and come on in the morning. There was little to be gained by scolding them, however, so grabbing the paraffin lantern I set off to search for the lame man. I followed the track for almost an hour but could find no sign of him. I imagined he must have got lost for in the blackened forest it was impossible to discern where the path lay. Abruptly I halted. An enormous python lay across my road, his head in the middle of the path. Cautiously I detoured round him and drew near for a better look. His sinuous body was as thick as a man's leg and stretched away out of sight into the dark jungle.

I had only been there a few seconds when the lame porter hobbled round the corner, still carrying his load of coconuts. What might have happened if I had arrived a few minutes later I hate to think for the poor fellow weighed no more than a hundred pounds and was well within the python's food range. Unarmed and crippled, he would have been easy meat.

As it was, I enlisted my companion's help to catch the snake. He grabbed its tail while I secured it about the neck and we soon had twelve feet of writhing serpent inside the sack previously occupied by the coconuts. Triumphantly we returned to camp where our prize took over both the home and name of Si-Ross, his forerunner having earned his freedom when I filmed him swimming in a nearby stream.

Chapter eleven

Old Rivals

It is not yet light but already the harsh, grating calls of the grey leaf-monkeys echo to and fro across the dewy *ladang*. Purba brings a pot of steaming rice porridge to the bamboo table and sets it before me with a mug of hot, black coffee. Nami still snores but the pigs are grunting in expectation of their morning feed. As dawn breaks I move off slowly along the main trail but soon leave by my own secret turning up the side of a narrow slope. A sleepy macaque gives a friendly threat, 'kra, kra,' as I pass beneath him. The tree-frogs continue bleeping, but the first crickets are warming up for the new day and several small birds sing tunefully from their misty perches. The hill is not steep but I take it gradually so as not to miss the important sounds that will tell me who lies to either side. At the top I sit on an old log, my ears straining for the slightest noise. A red-tailed tree-shrew scuttles about among the dead wood and fallen leaves but otherwise all is still and I carry on, following an ancient elephant path along the ridge. Wild pigs have sorted through the litter on the forest floor, ploughing up the soil, and I pause beneath a laden fruit tree to examine the discarded shells but they bear monkey, and not orang, teethmarks.

As the path improves I travel faster. Down the slope to my right an *Usoh* tree is coming into flower and the fragrant scent drifts pleasingly on the breeze. Humming busily, the early bees work hard to collect the sweet nectar before the sunbirds find the source. I cross the open ring where the argus pheasant dances and pass the mud-stained tree where the elephant likes to rub his back. Near the end of the ridge I slow down for I am approaching the main set of mineral caves where there is always something of interest to see. Sure enough a troop of leaf-monkeys are feeding below me. An old male spots me and cackles an alarm and the whole band leave with spectacular leaps and hurry away through the lush canopy. I make a note in my book then take up a position on a narrow ledge above the largest cave. I have an excellent view of the opposite hillside and have twice previously seen orangs from this point. I scan the slope through my binoculars but to-day there is no sign of life, no suspicious movements nor shaking branches.

A gorgeous black and scarlet minivet flashes about, flitting between the twigs and catching insects in mid-air or skilfully plucking them from their hiding places under the leaves. A bough stirs overhead and I look up to see a giant squirrel building his drey. The large, black and white rodent creeps stealthily along the branch, bends back a leafy shoot and cuts it neatly with his chisel-like teeth. Carrying his trophy in his mouth he hurries back to the nest and, with much rustling and bustling, struggles to fit the new piece into place. He surveys his handiwork and, satisfied, comes out to select another twig. The drey is a large, untidy affair, not unlike the nest of a young orang but rather more spherical.

While I sit and wait, a small, silvery lizard stalks but misses a fat fly and an orb spider repairs his worn web for another day's hunting. I leave my roost and climb down the hill to check the caves. A flurry of tiny bats swoop out as I enter and bend to examine the tracks on the sandy floor.

Although there are still signs of elephants none has been here since my last visit and only the deer slots are new. I wander down to the stream and watch a flying lizard angrily flashing his neck flag at a rival on a neighbouring tree. A young muntjak hears me coming and bounds away, white tail in the air and barking anxiously. Two long millipedes revel in the joys of procreation and an army of mindless termites march endlessly over a dead log, along a liana and stretch out of sight up the trunk of one of the forest giants.

I pause at the tinkling brook and pluck a broad, green leaf from an overhanging bush. I fold it twice, then open it out to make a beaker with which to scoop a few mouthfuls of water from the clear pool. Although I am not really thirsty, it is as well to drink while I can for midday may find me far from any stream. A dead mantis lies in the shallow water and white flatworms emerge from under the stones and swim like undulating ghosts to gather on the corpse. I am glad that I did not notice these tiny ghouls till after I had drunk but I have used these streams for many weeks and not suffered any ill effect.

The elephant track follows the course of the tributary before mounting the bank and sloping away up the hill. Another group of leaf-monkeys cry out a warning when they see me but, apart from one squealing infant, they are getting used to me now and stay to watch me pass. A family of gibbons begin to call at the head of the valley and stimulate another pair into action by the river; soon the forest resounds to their wailing chorus. Fruit shells beneath a *Dracontomelum* are certainly the work of an orang-utan but they are at least two days old and I am too late. I find a fresh nest, probably constructed by the culprit, and note its height and position. Passing a wild lemon bush I pluck a couple of young leaves and crush them between my fingers, savouring their fresh citrus perfume. On the next ridge I stop to rest and survey my surroundings.

10.45 Siamang group in Liang Djering valley start calling; gibbons below me still hooting.

11.00 Siamangs continue booming; gibbons have stopped. Occasional calls from male argus pheasant on hill 160° south.

11.08 Big band of long-tailed macaques cross ridge below me and pass into valley.

11.21 Family of red-crested partridges scurry past, searching for food beneath leaves.

11.25 Siamangs ceased calling. Calls from helmeted hornbill in same area.

I continue along Bear Ridge, so named because of the many claw-scarred trees on the summit. The day is hot and few animals are moving about.

12.05 Find fresh orang dung under fig and traces of feeding. No sign of animal. Collect dung for analysis.

I make a mental note to return to the fig tree next day as there is still plenty of fruit and my red friend may come back. I follow the track to the junction with the main trail for Kedai Ampuntuan and branch west to the pretty valley of Liang Djering. Shimmering, green-winged damsel flies flicker back and forth or perch immobile on the rocky boulders. I splash up the stream bed, letting the cold water soothe my tired feet. Dozens of tiny waterfalls twinkle where the droplets catch the sunlight and despite the midday heat the valley is cool and fresh.

I head towards the area where the siamangs were calling for, unlike the other animals that rest in the middle of the day, these elegant black apes show no decrease in activity, and although I have heard this particular family shouting on several occasions I have not yet seen them. I explore a new path leading up the steep hill and near the top is another fig tree where animals have fed. I suspect that both the siamang and cackling hornbills have visited the rich crop recently. I sit and wait.

13.25 Hear branches moving a short way down the slope. Investigate and find three siamangs. They are feeding on new leaves.
13.30 Still feeding. Male, female and large infant.
13.40 Female sees me and family rush off silently. Infant carried by male.

I turn my thoughts again to orang-utans and make my way slowly back to camp by a different route. I find no more signs of the red apes but one cannot always be successful. The joy of the forest lies in the wealth of animal and plant life that abounds there so that I discover something new almost every day.

The Ranun region supported a fantastic number and variety of monkeys. Every few minutes along the way I would meet long-tailed macaques, pig-tailed macaques or the pretty, silver-grey leaf-monkeys. Sometimes I encountered gibbons but these were more often heard than seen.

There were five resident families of siamangs. Two kept to the north of the river and the Liang Djering group I saw only rarely. The home family were a quiet trio frequenting the area round the waterfall but they had most interesting neighbours, a band of five who occupied the slopes behind the elephant caves. One enormous male, the largest siamang I have ever seen, seemed to lead the party and I christened him Hitam (black one).

One morning, while passing through their territory, I was alerted by violent sounds ahead and crept forward to witness two siamangs engaged in a fierce fight. To my surprise, Hitam was being chased all over the hillside by a smaller male I had taken to be his son. The animals flashed back and forth too fast for the eye to follow. Finally Hitam retreated and the younger ape returned to the rest of the family. A few days afterwards I came across Hitam, lurking near the other four but obviously keeping his distance. On several later occasions he was still alone and as far as I know he never rejoined the band. I have no idea what all the fuss was about; whether he had been kicked out of the family by his own son, supplanted by a handsome younger male, or was in fact a strange intruder only briefly tolerated by the rightful leader of the group.

Orang-utans were certainly not scarce and as long as the *Dracontomelum* season lasted I was able to clock up a good deal of interesting observation. One of the most fascinating characters was old Co. Co was the central figure of a group of some twenty orang-utans, who ranged over the three valleys of my research area. His great rival, Mo, was a huge male who ruled over a smaller band farther back in the hills. Whenever Co's company moved to the west of their range Mo and his friends seized their opportunity and sneaked down to the river. Like the Bornean antagonists these two argued their differences at long range, their wonderful booming roars reverberating across the valley. They also jealously guarded the river boundary against two challenging males on the far bank, with whom they were in constant dispute.

Mo had a round, friendly moon face and incredibly long hair on his arms and back, which gave him the appearance of a massive, woolly ball perched in a tree. Co, on the other hand, had a straggling beard, high forehead and his triangular face wore a haunting expression. Although he was very shy of me, he instilled much fear into the other animals and I saw one scared female hide in a mass of leafy branches for three hours to avoid Co, who was feeding below. Not until he went off to find somewhere to rest did she resume her own meal of sour fruit.

One morning I spotted Co travelling fast along the ground down an old elephant trail. I followed at a distance but soon lost all sight of him and my path brought me to the river. I never did find where he had vanished to but if I had not chased him I would have missed one of the most fantastic sights of the whole study. Beyond the torrent stood an enormous fig, whose pendulous fruits were just beginning to swell and ripen. A dark female orang had already discovered the feast and was working her way steadily through the topmost branches. On the steep hillside above me I found a gnarled root that made an excellent bench and with my cape and some boughs I was able to construct a flimsy, but effective, screen from which to spy on the ensuing activity.

For the first day things warmed up gradually. One by one other orangs found the juicy harvest and a succession of black faces appeared, ate their fill and moved on again. Towards evening a bunch of hornbills and a troop of long-tailed macaques joined in the celebration and the great tree swayed with their movement.

Next morning the orangs, two young males, two females and a juvenile, were breakfasting early. More hornbills and two giant squirrels kept them company. The orang-utans fed slowly and peacefully until the arrival of a small, dusky male. He mounted a long, sloping branch but somehow incurred the wrath of one of the larger males already in residence. The great bully turned on the newcomer and chased him, squealing, from the tree and across the hillside. The pursuit continued for some considerable distance with the two males careering round in a wide circle only yards apart. It seemed as though the pursuer would catch his quarry but suddenly he lost interest, their paths separated and the victorious male returned to the repast.

All was peaceful again but not for long. With wild leaping and much shaking of branches a family of five siamangs came hurtling through the canopy and halted opposite the fruit tree. Now, at last, I might see how siamangs and orangs got along together. The siamangs had obviously been attracted by the rich crop but would they dare to sample it when several large orang-utans were in splendid residence?

I did not have to wait long. With a mighty bound the male siamang leapt across into the lowest branch of the fig and was immediately followed by his family. Harrying the other diners they wove to and fro, skipping nimbly from one branch to the next and plucking the yellow fruit as they went. The elderly female was hampered by the tiny coal-black infant clinging to her lap but she easily outpaced the swiftest orang. Suddenly, without any warning, the four adult siamangs swooped in unison on a baby orang-utan who was playing happily by himself. His screams of fear brought his mother rushing to his defence and the evil black apes scattered. Furious, the red female gathered up her frightened baby and, hugging him fiercely to her, carried him out of the danger zone. As though upset by this incident, the orangs began to drift away until soon only one young male was left munching on the lowest bough. The siamangs were determined not to miss this chance and attacked again, four dark furies rushing down on the surprised orang. Clumsily he dashed to the end of the branch and with a hasty backward glance over his shoulder leaped into mid-air to avoid the vengeful terrors. He crashed into a tree below, landing badly, and hurried for a secure crotch to lick his wounds and wait for the hateful siamangs to leave.

The siamangs, however, took their time. When they had gorged themselves on the sweet figs they settled down to rest. The baby played alongside his mother but his movements were very unsteady and suggested he was only a few months old. I was amazed by what I had seen. I would never have suspected that a band of siamangs could put the much larger orangs to flight but by constant harassment and unexpectedly picking on young or solitary animals they had clearly won the day. This was one rival with which the Bornean red apes did not have to contend.

When I returned to camp I found we had visitors, Herman Rijksen and his beautiful blonde wife, Ans. Purba bustled about proffering mugs of lukewarm, muddy river water, and they were so exhausted after their trek through the midday heat that they accepted gratefully. Herman and Ans had arrived from Holland to work on behalf of the

World Wildlife Fund and the Netherlands Gunung Leuser Committee, which had been recently set up to help preserve the wildlife of North Sumatra. Herman's chief interest was the orang-utan problem and they had come to see how I was faring with the wild population. Confidently I told them that I would be able to show them a large group of our red friends. They were both surprised and delighted as they already knew of the unsociable habits of my Bornean animals and we awaited the morning with keen anticipation.

By the time we arrived at the blind a small, rust-coloured male had already settled in but he unobligingly wandered behind a screen of leaves so that we could not get a good view and after an hour deserted the tree. I was sure more orangs would arrive presently but in the meantime suggested that we should explore part of my research area. About a quarter of a mile along the elephant path I noticed a small sapling swaying suspiciously; only the weight of a heavy animal could cause such a movement. We stationed ourselves behind the convenient bulk of a huge trunk and shortly our patience was rewarded. A party of orangs were making their way steadily up the hill towards us. I was in no doubt as to their goal for nearby was a laden *Dracontomelum*.

In fact our caution had been unnecessary for the travellers were none other than my own particular favourites, a family of three. Karl was a handsome young male with a rather mournful, long face. His pretty bride, Kate, was large, pot-bellied and sported a bright red coat but her main claim to fame was as the proud mother of a sweet, little baby, Kim. Kim was about a year old and since he held as much interest for the male as for his mother I was certain Karl must be his father. Karl was a little shy and eyed us anxiously from his treetop seat but Kate and Kim were perfectly willing to show off their charms. They looked very pretty in the morning sunshine with Kim peeking out from beneath his mother's arm. Herman was soon clicking away madly with his camera while I hurried back to the hide to collect my own equipment. There were now two orangs feeding at the fig so we were in for a busy day. On my return I found Kate and Kim had climbed into the *Dracontomelum* and were picking hungrily at the round, green fruit. Kate tore off the thick rind with her teeth and rotated the flesh-covered stones between her mobile lips. When she had her mouth full she ground the stones round and round, rasping off the bitter-sweet segments and spitting out the hard kernels in a most unladylike fashion. Kim simply ambled along in her wake; the fruit were too tough for his baby teeth and he was still feeding mainly on his mother's milk. We took lots of photos and cine film of the co-operative duo until they settled down for a mid-morning nap. Leaving them to their peaceful slumbers we returned to my lookout by the fig tree.

Three males were taking a meal in happy proximity. Curiously enough the two who had participated in the hostile chase of the previous day now seemed to be the best of friends and had obviously forgotten their disagreement. Never in Borneo had I seen adult males tolerate the presence of others of their class but their Sumatran relations often showed such amicability at the dinner table. The trio were joined by a mother and her infant, the same baby that had been attacked by siamangs but apparently none the worse for this unpleasant experience. As the mother manœuvred past the old male he inspected her rear briefly but, not finding her to his fancy, he resumed eating.

The siamangs appeared later in the day but with so many big orangs about they were rather more respectful and managed to enjoy their dinner without any scraps. A neighbouring family of gibbons kept a close watch on the black apes, however, and not daring to approach the tempting feast, hooted with rage and frustration from the safety of a clump of *Itchap* trees. The unfortunate gibbons never did manage to snatch a snack in the five days the fig bore fruit but almost every other forest animal seemed to sample the succulent crop. Altogether fourteen different orangs made a total of sixty separate visits and the siamangs came three times to eat there. Other diners included long- and pig-tailed macaques, grey leaf-monkeys, giant and black squirrels repeatedly and Great, Rhinoceros, Pied, Black, Wreathed and Helmeted Hornbills all took their share.

Suddenly the female orang-utan, her infant and one of the males made a speedy exit from the tree. We looked about for the cause of this unseemly haste. A huge male orang was thundering along towards the fig and the branches quaked at his passing. The last remaining red ape almost flew out of the tree, just in time to avoid the new arrival, who settled down to a solitary feast apparently unaware of the alarm he had aroused. He was much bigger than any of our earlier acquaintances and sported a magnificent coat of long, shaggy hair. He was definitely the boss of this outfit and although the cheek flanges on his broad face were not as pronounced as those of Co and Mo he made a worthy rival. This must be the male I had often heard calling on the far bank of the river. The great ape lingered only an hour and with his departure the other orangs timidly returned. That day Herman and Ans saw a total of thirteen different wild orangs, a remarkable achievement for their first trip to the jungle. In spite of my nonchalance I, too, was thrilled at the great success of their visit, and it was with regret that I waved them off on the long walk back to the road where the Land-Rover was waiting to take them on to Kutatjane, their base for the next few months.

The number of visitors to the fig tree gradually dropped off and I never again witnessed anything so spectacular. Herman's trip, however, had created quite a reputation for my ability to lay on wild orang-utans for anyone keen enough to brave the long trail to my camp. When Mr Bangun Mulja, Conservation Officer for North Sumatra, arrived, Kate and Karl gave a delightful performance, and an American, Ben Grey, accompanied me on a long trip to find old Mo and a female with two youngsters. The Forestry Officer from Sidegalang was less fortunate but seemed well pleased by the consoling sight of a family of elephants.

News of my success finally came to the ears of one of the local *pawangs*, a caste of men found throughout North Sumatra. These are the magicians of the jungle. Hailing from the surrounds of Blangkedjeren in Atjeh, these professionals are trained under famous mentors in the arts of jungle lore and magic. They make a living as guides in the forest or from practising their medical skills and other secret talents. They claim to be able to communicate with the animals and hence call them up at will. More unpleasant is their active encouragement of superstition in the *kampongs* so that many are able to earn a reasonable living by liberating affected souls from the spells of their colleagues.

Each wizard specialises on a different creature so that there are *pawang rusa* (deer), *pawang gadjah* (elephant) and even *pawang mawas* (orang-utan) and it is to no avail to try and gain the help of any old *pawang* if your rice crops are being trampled by elephants. The correct specialist must be called in. I had refused the services of such men with the joking claim that I was a *pawang kupo-kupo* (butterfly magician) myself. Now it seemed that my success with my red apes and some fortunate cures of ill *kampong* folk were threatening the profitable business and, on the pretence of a fishing trip, the local *pawang rusa* paid a visit to Samporan. He was much interested in an Indian puzzle ring that I wore and was sure that herein lay my power. The curious fact is that later that night when I was bathing I noticed that the ring was no longer on my finger and I never found it again. I could not help wondering if he had spirited it away. Even more disturbing, from then on orangs were no longer so easy to find.

Chapter twelve

Love in the Treetops

The orang-utan has always been painted as a creature of excessive libido and there are countless tales of night-long orgies in leafy nests, abducted Dyak maidens, passes made at zoo-keepers and orangs kept for sexual entertainment.

Young animals in captivity are, indeed, sexually precocious and mischievous, constantly masturbating and indulging in erotic games. But wild orangs, despite their reputation, are comparatively restrained in their love-making. In fact, chimpanzees would fit much better with our image of bestial sensuality. Like many terrestrial monkeys, female chimpanzees develop swollen, pink posteriors when they are in oestrus and show little discrimination with whom they couple. Hence, males and females can enjoy intercourse many times a day and sex has become little more than a greeting, an intimate handshake.

Perhaps it is because of the dangers involved in bouts of wild abandon high in the forest canopy that sexual behaviour is so infrequent in the treetop apes, the gibbons and orang-utan. Often I watched as orang boy met orang girl but they passed each other by without showing the slightest interest.

My first insight into the love-life of the orang-utan came in the winter of 1969 in Sabah. I had spent several rewarding days following a large, dark female, Ruby, and her juvenile son, Richard. Unexpectedly, on our fourth day together, they hurried back the way they had just come. I heard high-pitched screams ahead and rushed up to see what was happening. By the time I reached the scene the crying had ceased and so I was none the wiser. I found, however, that there were now three orangs, all resting quietly. Had Ruby known that the other animal, a sub-adult male called Humphrey, was up the slope or had she met him by chance? I could not tell but the trio stayed together throughout the day. So long as Humphrey kept his distance Ruby seemed quite at ease but she quickly moved out of range whenever he tried to be more intimate. That evening Ruby and Richard shared a nest at the very top of a tall tree and Humphrey slept far beneath them.

When I returned next morning Ruby and Richard were still abed and there was no sign of Humphrey. The day wore on. Richard played among the branches but Ruby emerged only briefly to chew a few leaves before returning to her springy couch. It was unusual for orangs to sleep so high and even stranger that they should stay by their nest all morning. I wondered whether Ruby could be sick. At half past two her strange behaviour was explained. Humphrey reappeared and climbed towards the lethargic pair. Richard began to cry and hurried to his mother. Ruby's face grimaced with fear as Humphrey seized her from behind and dragged her out of the nest. The ardent lover bit and struck her then, clasping his feet firmly round her waist, proceeded to rape the unfortunate female. Ruby struggled furiously and scrambled slowly and awkwardly downwards. Richard clung to her side screaming. With his tiny fist he pummelled the larger male on the chest. Not at all discouraged, the villainous Humphrey lowered himself arm over arm, still keeping Ruby firmly in his grasp. Ten minutes later the struggling group reached the ground and broke up. The protagonists began feeding once more as though nothing had happened. They nested close together again that night but next day Humphrey parted from the little family and continued on his travels.

I saw several similar matings in Borneo – brief, rather violent assaults on unwilling females. I found it hard to believe that this could be the accepted norm for orang-utan life. Females were obviously terrified by these attacks and often went out of their way to avoid approaching males. Such a state of affairs could hardly be conducive to procreation. Moreover, what were consorting couples all about? Many times I had found males and females travelling together in peaceful harmony. Surely these were the reproducing members of the population; yet their modesty prevented me from being certain. Perhaps their lack of action was a reflection of the noticeable drop in birthrate in the Segama area.

Since 1968 there had been a steady decline in the number of infants north of the river and in 1970 no new babies were born there. The proportion of males engaged in consortships was only half as great as in the population south of the Segama but they were calling almost three times as frequently. This halt in breeding and the males' increased antagonism seemed to be in response to overcrowding, for this was the area disturbed by the presence of wandering, homeless orang-utans that had been displaced by timber-felling operations farther north. The orangs seemed to be refraining from reproducing at an unsuitable time.

Many other animals have natural methods of regulating their numbers. Experiments have shown that rats and voles when they are overcrowded and suffering from social stress will stop breeding. They become aggressive, even cannibalistic, and show reduced fertility and higher rates of abortion and infant mortality. The method employed by the orang-utan, however, is rather different with the increased aggressiveness of the males exhibited in long-range calling and a reduced tendency to consort with females for mating.

Such a capacity to affect birthrate by changes in mating behaviour, a form of natural birth control, is perhaps unique among animals.

Still not satisfied with my scanty observations on this important phase of the ape's life, I decided to turn my attention to captive orang-utans to see if their behaviour would cast more light on the matter. British zoos are well stocked with orangs so that I had no difficulty in choosing subjects. Regent's Park proved most rewarding. Boy shared a cage with females Twiggy and Bulu, while next door a younger male, Gambar, lived with Blossom and Bunty.

I seemed to have hit on a lucky period. In three days I saw several sex-bouts from the four possible partnerships. Sumatran Gambar strutted round his quarters, chest out and arms held straight by his side. Little Blossom approached him playfully but he had eyes only for the elusive Bunty. Nimbly she slipped aside as he drew near, but eventually she acquiesced and joined him in long sessions of play; poking, tickling and biting with Gambar frequently mounting her.

Visiting mothers hurried their puzzled children on to the next cage only to be confronted by Boy being equally immodest with one of his mates. Boy was a tall sub-adult male just beginning to fill out into the giant he will no doubt become. His most co-operative partner was Bulu, now an adult female but once in the newspaper headlines as the first orang-utan born in the zoo. Keen though Bulu seemed for Boy's affection, it was Twiggy who interested him most. Twiggy, despite her name, was enormously fat and surely pregnant. She obviously felt that she had seen quite enough of Boy for she responded to his amorous overtures by roughly biting and hitting him. Typical of his sex, however, he persisted until he finally got his own way and the couple completed their love-making on the straw-piled ledge.

I was not surprised to learn a few months later that

Twiggy, Bulu and Bunty all had babies. Sadly Twiggy's baby died and Bulu showed so little idea of mothercraft that the zoo-keeper's wife had to relieve her of her maternal duties. Bunty, however, who had been born in the wild, proved to be a perfect mother and was allowed to keep her Sumatran–Bornean hybrid infant.

Under captive conditions, with no natural outlets for travel and foraging, these animals undoubtedly diverted more time and energy into their sexual activities than their wild counterparts. I was, nevertheless, greatly impressed by the relative peacefulness of the proceedings and the females' willing participation in, and obvious enjoyment of, the sex-play and mating. This was very different from the rapes I had witnessed in Borneo. The relationships between Gambar and Bunty and Boy and Twiggy were so similar to those of the consorting pairs I had met in the wild that I was convinced that it was within such stable consortships, rather than from rape, that most wild babies were conceived.

My Sumatran findings confirmed this absolutely. The first orang-utans we had found in Langkat had seemed to share a contented partnership and the female, after a preliminary show of coyness, had joined willingly enough in their mating.

I encountered only one other Sumatran couple indulging in sexual behaviour. These were my old friends Kate and Karl. On the first occasion I met them Karl was pursuing Kate through the canopy. She squealed and struck out at him, just as Twiggy had treated Boy, but finally gave in to his persuasion, wrestling and grappling in the treetops. Tiny Kim clung on to his mother's shaggy coat but these cavortings were nothing new and he exhibited no alarm at his parents' mad performance. Kate and Karl seemed very happy together and showed no sign of breaking up in the five months I knew them. Karl was an attentive mate, never slow to show his feelings for Kate. With one arm draped lovingly round her shoulder he would play with baby Kim.

Sometimes his attention would turn to tickling and biting, encouraging his partner to participate in more bouts of struggling and chasing. When little Kim felt too left out of things he whimpered and hurried to his mother to be reassured by a kiss from either parent.

With a more definite knowledge of the orang-utans' family affairs I could at last build up a picture of their life history and development from birth to maturity.

During his first year the baby orang-utan is in almost permanent bodily contact with his mother, clinging tightly to her side as she travels, or clambering over her in an exploratory fashion when she pauses to feed or rest. Although these babies see, and often sample, the foods their parent finds in the forest, they are nourished almost entirely on her milk. They are groomed by their mothers and share her nest at night. If an elder sibling or an attentive male is accompanying the family they may sometimes entertain the youngster but usually he must rely on his mother as his sole playmate.

By his second year the infant is less dependent. Although he never strays far from his mother's side he rarely needs to be carried and is capable of finding some of his own food. At night he still sleeps with the female but during the daytime he experiments with building play-nests and sometimes tries out his efforts for a short rest. He will now join willingly in rough games with any other orang who will tolerate him and by his third year is, like Monty, actively searching for playmates, even if it means deserting the family for a few hours.

Juveniles remain with the mother and may show infantile behaviour, such as sleeping with, and occasionally even suckling and riding on, their parent until the birth of a new baby. From then on the close bond between mother and offspring starts to break down. The juveniles, particularly the males, gradually wander farther and farther away from the family and spend increasing periods alone. Young females are less adventurous and often stay with the mother and play with the new infant. It is probably at this time that they learn the fundamentals of baby care since by the time they produce offspring of their own they will be leading a solitary life with no older females around to turn to for example or help.

By the adolescent stage few animals remain in their parents' company. They prefer to travel alone or perhaps link up with some other youngster in similar straits. Throughout adolescence and sub-adulthood orang-utans are busy establishing their own ranges and learning their place among those neighbours they encounter. It is now that they become involved in their first consortships. Though still not fully grown, females can breed at about seven years old. They enjoy only a brief period of solitary freedom for, from the time of their first pregnancy, they are continually involved in motherhood with a succession of new infants at two- or three-year intervals.

Males continue to grow until well into their teens, when the large face flanges, heavy jowls and long hair of high rank develop and they begin calling. If they find suitable and willing mates, they engage in consortships. Like Karl's these may be long-lasting, but often they are brief affairs and in between times they lead a solitary life, joining in the power struggle with their rivals for rank and dominance. The older patriarchs seem so preoccupied with calling and status that they have retired completely from family life, leaving the responsibility of procreation to their juniors. Even without such competition, however, the young males do not have an easy task since the female orangs require considerable persuasion that this is the right mate for them.

Co and most of his band had crossed the narrow ridge into the valley of Liang Djering but Tom, a young male, could not bear to leave and alternated his meals between a *Dracontomelum* and a *Kuagee* tree about thirty yards away. Towards evening he was feasting in the latter when an adolescent female arrived to sample the *Dracontomelum*. Tom eyed her interestedly but, surprisingly, made no attempt to interfere with the trespasser. He maintained a polite distance while she dined, but when she was still there eating his precious fruit in the morning it was too much for him. As quietly as he could he stole up on the unsuspecting culprit and I thought he might cuff her soundly for her daring. Instead he settled beside her with a friendly grunt and both animals continued feeding.

This peaceful scene did not last long. Tom decided to speed things up and sidled across to better their acquaintance, but when he tried to stroke her she squealed in dismay and took to her heels. Undeterred, Tom swung after her and this time managed to catch hold of his coy companion. He pulled her, still squeaking, into an old nest where the couple tumbled to and fro, biting and tickling playfully. Just for a moment Tom let go and his precious find slipped away and crossed into the next tree. The young lady's bashfulness seemed to stir Tom to further displays of ardour and he paused in his pursuit to show her what he was made of. For well over a minute he indulged in a weird performance. Arms and legs akimbo he swung a dead, leafless sapling round and round in circles, all the while moving his chin slowly from side to side. The long hair on his arms swayed hypnotically back and forth with the motion. It was a truly wonderful sight and seemed to have the desired effect on his intended. He hurried off after his new love, who ran just slowly enough to let him catch her again and soon they were rolling on another leafy couch, gurgling and wrestling. For the rest of the day they fed side by side in a mass of vines and that night they both nested high in the same tree. I hoped this might be the start of a beautiful relationship but somehow they didn't quite hit it off and when I arrived at dawn Tom's shy girl-friend had given him the slip.

Chapter thirteen

Jungle Honeymoon

I was reminded that my own bachelor life was about to end when I received a letter from Kathy, my fiancée, who was at last heading east. I hurried back to Medan and spent a hectic day obtaining an exit and re-entry permit, visiting government offices in all four corners of the town to pay the required, but inexplicable, taxes. I flew into Singapore just in time to meet Kathy's plane from England but as she was almost the last of the two hundred passengers to enter the arrival lounge I had almost given up hope when she rushed to greet me. To my horror she was burdened with white dress, iced cake and all. To match this terrifying respectability I had to buy a suit, shoes, shirt and tie before we could return to Sumatra, where Len Hanham, the British Consul, had been arranging matters on our behalf.

The Consul's Chinese secretary had gone to great lengths to ensure that the town registrar would be in his office on time to legalise the marriage and Kathy gaily swore away her single status in an Indonesian ceremony she didn't understand. Medan's small European community and my Indonesian friends saw us safely re-wed English style at the local Methodist church before we retired for champagne at the Consulate. Later there was great feasting at the home of

Nikki and Michael Jenkins and we all overindulged ourselves on the *reistafel*, which the servants had been preparing for the last two days. Nikki is a fanatical animal saver and is unable to pass any badly treated creature without purchasing it from its owner. Consequently the garden was full of an astonishing array of parrots, rescued cats and spared chickens. Seamus, the orang-utan, looked very sweet kissing the bride even if his teethmarks did remain visible for several weeks. He had been shampooed specially for the occasion but behaved horribly, stealing wedding cake, chasing the cats and biting guests' ankles.

Like every good town Medan had a resident MacKinnon and, like every good MacKinnon, Ed had a set of bagpipes. To the amusement of the Indonesian *betcha* boys peering over the garden wall, Ed marched back and forth with his wailing pipes drowning the shrieking of the parrots. To the raucous strains of 'MacKinnon's Lament' the cake was sliced with a ceremonial Malayan *kris* and rum punch imbibed from silver goblets. It was a splendid occasion.

In high spirits we were whisked away by the Rijksen's Land-Rover, the World Wildlife panda emblazoned on the door. We careered madly through the town, the driver roaring with mirth at the strings of tin cans jangling along behind us and the consternation they caused among the locals, who all turned out to watch this latest crazy escapade of the mad Europeans. The sick baby orang-utan Ans was nursing must have had a most uncomfortable journey. They had found him half-starved in a *kampong*; the poor little creature was much too young to be separated from his mother and would certainly have pined away on the proffered diet of rice if Herman and Ans had not intervened. As it was he was far from happy and it needed all Herman's veterinary skill to keep him alive.

Kathy and I, too, had charge of a young animal, a baby leopard cat, travelling in a *rotan* picnic basket. We had found the animal for sale at a roadside stall and been quite

The flying lemur – a zoological mystery. Below, a flying snake devours his prize

Above Old Co, and Kate with Kim
Opposite Karl
Overleaf Siamang and, below, the author

unable to resist the attractive creature, who charmed us with her kittenish ways. With her white face stripes and easy grace she looked like a perfect miniature of a tiger and she was duly christened Tigger after her lordly cousin. Certainly the deep growls that came from the little basket were worthy of a much larger animal.

We wound on through the mountains with occasional stops to purchase delicious mangosteens and snake-skinned *zalak* fruits from roadside stalls. After a long, hot journey we left the main road to rattle through a tiny *kampong* of wooden shacks with palm-frond roofs and, narrowly missing a pair of browsing water buffalo, screeched to a halt in the village of Lau Ranun.

Even a Land-Rover could go no farther, so Herman and Ans waved us goodbye and dashed off again on the road to Atjeh, their month's supply of Coke bottles jingling wildly as the vehicle jumped along the rutted track. A crowd soon gathered to inspect Kathy and all our luggage and there did seem to be an awful lot of it. I reckoned that we should get three porters to help carry the pile to my camp, a nine-mile walk away. Unfortunately, none of the villagers was any too keen to spend the night at my humble abode and they demanded a thousand rupiahs (equals £1) each for the journey, some four times the normal rate for these parts. I was not going to be bulldozed in this manner and defiantly declared that we could manage ourselves. Heavily laden, we staggered off with most of the village following to enjoy this amusing spectacle.

It took us over three hours to reach Lau Djohar and Kathy has never quite forgiven me for that frightful journey. Under a blazing sun we toiled and sweated up and down steep hills, through muddy *padi* fields and insect-infested banana plantations. Several times Kathy swore she could go no farther. Squatting exhausted on an enormous heap of our goods and chattels, she could only mutter that it was a great pity we hadn't made this trip before we were married, then she might have known what she was letting herself in for. When she realised that the porters had only requested a pound each she was furious. She would cheerfully have paid them double that amount to spare herself this ordeal.

At last we emerged from a coconut grove overlooking the Ranun River. As we made our way down the meandering path a small boy dashed out of a nearby shack and hauled a dugout canoe from the shelter of the bank. We boarded this precarious craft and our little boatman poled furiously to ferry us across the sluggish stream.

A narrow track on the far bank led us between straggling oil palms into the *kampong*. Here Nami's son, Pangsa, appeared to greet us and within seconds we were comfortably seated on two cane chairs spirited out from some hidden corner of the village. Pangsa struggled with our luggage while his brother hurried off to procure fresh coconuts, real nectar to a thirsty traveller. These were not the wizened, old offerings much prized at English fairgrounds. Gratefully we drank the cool, sweet milk and scooped out the thin shell of gelatinous flesh. A pot of rice was soon bubbling and Nami's wife bent over the fire preparing the evening meal. Every few moments a new influx of visitors mounted the short wooden ladder into the dim hut and the womenfolk were kept busy proffering endless cups of tea. The local policeman, Purba's uncle, was treated with particular respect and a space hastily cleared for him as near to us as possible. By now the tiny, smoke-filled room was so crowded that it was difficult to breathe and we were glad to pick our way through a mass of bodies into the fresh night air. We strolled down to the river to bathe, shocking the natives by this simple action. They have separate washing places for men and women, a custom which sometimes leads to the amusing sight of the two sexes bathing in irrigation ditches on opposite sides of the road, yelling gaily to each other and at the traffic roaring past between.

Refreshed and cooler we returned to the house hoping that most of our curious visitors would have taken the hint and retired homewards. But if anything there were even more people packed in; even the tiniest baby had turned out for the evening's entertainment. We realised to our horror that, far from planning to go home to let us repair to bed, it was to view this spectacle that they had come. Without so much as a sheet between us and to a chorus of giggles and nudges we were forced to settle down on a new palm mat, specially laid over the hard boards. We lay there, eyes firmly closed and stiff with embarrassment and I fear our audience were sadly disappointed for soon they began to drift away. The more hopeful among them reappeared at daybreak but we had anticipated this manœuvre and were up well before dawn. It had been a restless night punctuated with barking dogs and crying babies, while Pangsa and his mother sorted through a large pile of *kanari* nuts at our feet.

Before the morning dew had dried we were off again but this time Pangsa and two stalwart youths were bearing most of the weight. We wound on through the forest, past the hot springs, up and down the multitude of small hills and finally rounded the last bend to my forest camp. Nami and Purba were wreathed in smiles as they greeted us and I introduced them to Kathy.

We released Tigger and anxiously awaited her reaction to the new surroundings. We need not have worried. Within minutes she had asserted her rights of territory and trotted about scattering chickens and rebuffing the friendly pig. Before many days had passed she had reduced the hen, a most conscientious and belligerent mother afraid of no man, to a harassed and careworn bird desperately marshalling her few remaining chicks into the safety of their basket.

To honour our wedding and Kathy's arrival Purba decided to kill a pig, a great sacrifice on his part as he regarded them both as personal friends. While he set about this gruesome task I gave Kathy a brief tour of the area showing off the giant squirrels, flying lizards and our resident leaf-monkeys. Hungry after our trek, we turned up for dinner with visions of succulent roasted pork. Unhappily for us Purba had decided to lay on one of his specialities. From the resulting dish it seemed as though he had chopped the slaughtered pig into little pieces and liberally laced these with water and cinnamon bark to produce a very dubious stew. Grinning with delight and self-satisfaction, he piled our plates high with rice and a generous helping of the dish. Kathy's face was a picture; I am sure she would have been thankful to forget her zoological training and remain in happy ignorance of the choice bits of anatomy offered to her. Bravely she choked her way through the first bowlful but when the same concoction reappeared every mealtime for four days, and each time a little higher, she had to admit defeat. Finally she begged Purba to make her some *bubor* (rice porridge), a much inferior dish, and he was so concerned at this obvious sign of ill health that he packed Turut off to find bananas for the *mem* with dire threats of what would happen should he return empty-handed. I must admit that I was glad to see the end of the pig but with a beaming Purba standing over me, a heaped spoon ready for refills, it wasn't easy to refuse his culinary efforts. After this disastrous celebration we agreed to spare the remaining pig and Purba would take it off for his own wedding in the near future. This later proved to be a most unfortunate decision.

Our own contribution to the nuptial feast consisted of some rather travel-worn wedding cake and a bottle of excellent wine brought at considerable expense from Singapore. The men sipped warily at this doubtful beverage and would obviously have preferred a bottle of the local fiery brandy, an excellent bargain at only one hundred rupiahs (two English shillings). After catching up on the latest news and attending to the welfare of the various animals in my menagerie we were ready for bed. The sleeping arrangements were scarcely more satisfactory than the previous night. Although we did have a narrow mattress we had to share our open platform with six inquisitive Indonesians: my own three helpers and the porters who had stayed for the treat of Purba's cuisine. Far into the night they chatted and smoked and whenever we succeeded in extinguishing one lamp they hastily relit another, determined that no wandering spirit should creep up on them unawares.

Sleep when it did come was light, troubled and short-lived. For who should come prowling round but Musang, the civet, attracted by a leg of pork hanging from the rafters. He made short work of this tasty morsel then looked about for further fun. Musang was quite a trial. I had found him as a baby and released him at our camp. Immediately he showed his independence by finding and devouring a nest of young rats in the roof. He was an accomplished climber and vanished high among the trees each morning but when darkness fell he returned to pester us and cause havoc in the foodstore. His toilet habits were far from clean and he was a most unwelcome sleeping companion since he couldn't resist giving a friendly nip to any fingers or toes protruding from under the blankets. To-night he smelt Tigger, contentedly curled up between us, and pounced on the dreaming kitten. Shrieking madly Tigger spat and scratched at the invader and in the resulting fracas Kathy and I suffered more damage than either of the combatants. The wrestling match was repeated several times that night and every night to come until we finally left Samporan.

There were further disruptions at breakfast when Nami let the tree-shrew escape. We saw a flash of red, a flick of a bushy tail and the squirrel-like creature disappeared among the undergrowth. I didn't blame him for making a bid for freedom. He shared a large mesh cage with two slow lorises, Peter and Kukang. The arrangement should have worked perfectly satisfactorily as the tree-shrew was active during the day and the lorises didn't wake up till night-time. If he delayed his bedtime, however, to linger over a choice slice of banana, then his neighbours took great delight in ambushing him on the way home to his hollow log. Although they were slow lorises, and their speed amply justified their name, they showed sly cunning in their tactics and took a vindictive pleasure in taunting the unfortunate tupaiid.

Tree-shrews closely resemble small squirrels but have long, pointed noses. Zoologists argue over whether or not they should be included in the order primates with the monkeys and man, but the tree-shrews themselves are quite happy as what they are – nimble, forest insectivores, marking out their territories by rubbing landmarks with their scent glands. Ours had fallen victim to one of Turut's snares and I was glad of the opportunity to take a closer look at the shy creature. My first surprise came when he refused to eat fruit or rice. Even the tigers in the local zoo ate rice! We did eventually persuade him to try bananas but for nearly a month he kept us constantly searching for insects to satisfy his insatiable appetite. Beetles, grubs, crickets, grasshoppers and moths were all equally acceptable. Tall and motionless, he would stand in his corner, his lower jaw working comically from side to side as he crunched on these delicacies. Given the chance he could eat his own weight of insects at a single meal and we all sighed with relief when he escaped.

The lorises were easier to please. They, too, enjoyed fat, green crickets but they were quite content with rice, bananas and milk. In appearance they were the most delightful of animals, something like a cross between a bush baby and a teddy bear. They could easily pass for cuddly toys as they slowly rolled their heads from side to side, quizzically inspecting the world. Lorises are included among the primates without argument. Their big eyes, situated on the front of the face, give them the dual advantages of nocturnal and binocular vision. Lorises have finger-nails rather than claws and their hands and feet are beautifully adapted for grasping twigs and clambering slowly through the treetops. Instead of a tail they have a mere fluffy apology for they have no need of such an appendage.

On emerging each evening Peter and Kukang spent a long time licking their fur and scratching with the curious, long grooming claw of the first toe. Peter was a fat, mature male who had been caught in the act of ravaging a villager's pawpaw tree. When he arrived Kukang was so tiny that he fitted comfortably into my shirt pocket. We unearthed a medicine dropper and dried milk and Purba played the part of a rather unwilling nanny. Very anxiously we introduced the infant to Peter. The pair joined in a curious creaking chorus then Peter seized the poor baby in his big paws and began to lick him vigorously. Peter made an excellent foster-mother and looked very sweet with Kukang clinging to his manly chest. He had no objection to Kukang receiving milk from his pipette provided that he, too, could join in this sweet luxury. Very soon Kukang was able to travel around the cage alone.

For such slow movers they were remarkably adept at catching crickets, not an easy task, as we found when collecting provisions for our pets. It was as though the insects saw the furry bundles advancing but thought, 'Oh, there's a good five minutes yet before these sluggards will act,' and so were trapped by their own lack of caution.

Like the tree-shrew the lorises marked out their domain but with a trail of urine. They looked very funny creeping along with haunches lowered, drawing out a thin stream along their branches and all over the roof of the tree-shrew's nest box.

Leaving an apologetic Nami to clean out the animals, Kathy and I left camp to see what had been happening in my absence. She was very keen to see my orang-utans but I feared this would not be easy with the animals so far away. We wandered far over the ridges and it was a great pleasure to show off my little kingdom and favourite haunts. At the Elephant Caves there were fresh tusk marks but we saw no other sign of the great beasts, much to Kathy's relief for, when I explained that one only had to climb a tree to avoid a charge, she pointed helplessly at the smooth trunks rising up for tens of feet without any foothold. We splashed down cool mountain streams and bathed in clear pools. Together we marvelled at weird, twisting fig roots, enormous buttresses and exotic tree-ferns, each supporting its own world of animal life in the rainwater trapped among the plume of leaves. Leeches looped towards us, a horrible but fascinating sight, and Kathy was initiated into their blood-letting in spite of her careful precautions.

I had lived in tropical rainforest for so long that I had come to take the wealth of animals for granted. I had forgotten the frustrations of an English wood and sitting out night after night, prey to every passing midge, in the hope of seeing a cautious badger emerge from his sett. Kathy was amazed and delighted by all the creatures we met: tumbling monkeys, snorting pigs, timid mousedeer, a writhing mass of millipedes. Round every corner another actor laid on a wonderful display. Even the local siamang family visited the hill above camp and went through their paces, calling to impress the new visitor. But my orang-utans remained elusive and it was several days before we heard Co roaring in the distant hills.

Kathy was particularly interested in the many species of tropical squirrels as she was studying the roguish grey squirrel in Oxford for her own D.Phil. The Ranun area supported a large selection of these rodents in all colours and sizes. Tiny pigmy squirrels bustled about, nibbling delicately on bark, their jerky movements giving an impression of cine film run at the wrong speed. Higher in the canopy their larger cousins all had their appointed roles and ranges. There were nondescript brown *batchings* with narrow side stripes, a big group of near relatives only distinguishable on close examination by differences in size. Weaving stealthily through the branches, the dark horse-tailed squirrel enjoyed a rich diet of acorns, figs, winged dipterocarp fruit and the famous durian. He was very common in coconut plantations, where he feasted on the nuts and earned the hatred of the villagers.

Ratufa, the giant squirrel, was another fruit-eater and if one sat long enough by a fruiting fig eventually he would arrive to join in the spoils. His alarm display was an incredible sight; he flagged his long tail so vigorously that two waves of undulation moved along it at once. I have seen as many as four *Ratufa* together hurtling round a trunk in a wild game of catch-me-if-you-can, which left me quite dizzy. *Ratufa* was not a quiet animal; he crashed about, scattering distinctive large, leafy nests throughout the treetops. Although Prevost's squirrel is not at all common in Sumatra we spotted one idling near a mineral cave. He was a handsome fellow with a black back, white belly and red waistcoat. But our favourites were the flying squirrels, who emerged when their less well-endowed cousins were thinking of bed. There were squirrels everywhere; in the trees, in the air and on the ground.

Hot and sticky from our long daily trips, we would grab our wash things and thankfully dash down to the waterfall. It was an idyllic spot, screened by a thick curtain of bamboo and with only an occasional wandering fisherman casting his throw-net to disturb our privacy. As we scurried across a carpet of pink-flowered herbs they shuddered and closed their leaves. These were sensitive plants and tall bushes of the same family straggled along the beach; if one brushed against them the whole shrub trembled and collapsed.

We plunged into the icy water and swam out to a shallow sand bar. This was the best part of the day. The cool, muddy river washed away the dust and heat of our travels and soothed our smarting scratches and leech bites. Relaxed and feeling really clean, we lay on the smooth rock slabs, drying ourselves in the fading sunlight and absorbing the beauty of our surroundings. A medley of gorgeous butterflies swirled and fluttered against a backdrop of shimmering spray over the plunging falls. Blue, green, turquoise, scarlet, orange, yellow and white, looping the loop, tossing and turning in a shower of confetti that slowly drifted on to the beach. They sipped daintily at the sulphur spring, patterning the pool with gay whirls of colour. At our approach they rose as a cloud, spiralled and sank to rest again. Silhouetted against the darkening sky, a family of monkeys scrambled to their sleeping posts. At peace, we lingered until the last rays of the dying sun and the rising clamour of the jungle night warned us that it was time to return to camp for our evening meal.

Although we had found a superabundance of other animals during our fortnight's honeymoon, I had almost given up hope of discovering the orang-utans. We had set off in the early morning on a long farewell trip through my study area. After following the river for a couple of miles we climbed up one of the main ridges and headed into the hills. By midday we had reached the Limestone Lookout. Leaping from pinnacle to pinnacle of the sharp eroded rock, the pitted slabs wobbling beneath us, we collected orchids and took photos of the amazing view. Far below two hornbills glided down the valley of Liang Djering, their outstretched pinions glinting in the sun. We waved goodbye to the distant rolling hills and slithered down from our precarious vantage point. Roused from his sunbathing, a four foot long monitor lizard blundered away as we hurried along the dry ridge to the welcome enveloping shade of the jungle.

We had been rambling for about six hours and were hot and tired; soon we would have to turn homewards. Suddenly I smelt orang-utans. On we tracked, carefully scanning the wall of greenery for a leafy nest or some other encouraging sign. A blurred red shape caught my eye and there they were within yards of us – a mother and her youngster, like two hairy rugs sprawled on a branch. Only their flickering eyes betrayed any sign of life. None of us moved. We stared at the orangs and they gazed dolefully back. At last mother decided she had suffered sufficient intrusion and, without any sign of haste, swung off downhill, the juvenile following close at her heels. Not caring how much noise we made we scrambled after them, sliding down the slippery slope, until they vanished from sight. I was delighted that at the last moment my 'magic' had not failed me and I had been able to find orangs for Kathy.

Although there were still three weeks before we must return to Oxford to write up our respective theses, I had to visit a few other places to make some final observations. The next day we left Samporan. Before our departure we had to find homes for the members of my menagerie. The tortoise was no problem. He hastily withdrew into his spiny shell at the unaccustomed attention but was soon burrowing happily into a pile of crackling, dry leaves. Si-Ross, the python, also made a quick return into the thickets from whence he came. He was quite undeterred by the noise and display the mother chicken made when she saw him emerge from his box. Peter and Kukang were set free on our last night. They presented an incredible sight, trundling across the ground at considerable speed, their tailless bottoms waggling from side to side. In no time they had climbed to the very top of a tall *Usoh* tree and sat watching us, their bright eyes shining back in the torchlight.

We were up at first light to persuade a resisting Tigger into her basket and an even more unwilling Musang into a wire cage. Forgetting our past differences we had decided to move the civet well away from camp since later visitors would soon put an end to his evil tricks. I was leaving my hut and most of the household paraphernalia to Turut so that we would have less to carry, or so I thought. I had bargained without the careful Indonesians. They conscientiously collected every discarded tin and bottle and crammed them into sacks that were already overflowing; all would have their uses in the *kampong*. Helplessly we watched our pile of luggage double in size. Extra porters had been engaged for the exodus but even so we set off sadly overloaded.

Stopping regularly for rests and at one clearing to release the wicked Musang, we made our way slowly back to Lau Djohar. One of the porters who had been carrying Purba's pig collapsed here and refused to carry anything more, so we were even more laden for the last stages of the journey but, at last after seven hours of weary progress, we reached Lau Baleng. Losing half our belongings in traditional exchanges of gifts, we bade our farewells to Nami, Purba and the Pardede family just in time to catch the last bus north to Kutatjane, where we were to visit the Rijksens and see how their conservation plans were getting on.

Chapter fourteen

Apes in Danger

Compared to our own grass hut the Rijksen home was a palace. Among its many attributes it boasted a clean *mandi*, insect-proofed windows, books and chairs and, luxury of luxury, a gleaming fridge full of cold drinks. We sat in the comfort of the living room, sipping Coca-Cola and relishing the almost forgotten tastes of mangosteens and *rambutans* as we exchanged our news. When darkness fell Kathy and I wandered outside and suffered the ravages of the Kutatjane mosquitoes while we watched the great flying squirrels gliding effortlessly back and forth between the durians at the bottom of the garden.

Grazing placidly on the lawn was an unusual-looking, black, woolly beast. We approached the blurred shape and discovered it was a rare serow or wild mountain goat, known only from South-East Asia and the Himalayas. But what on earth was it doing among the Rijksen flowerbeds? It transpired that Herman had received a report that two serows had been captured near Blangkedjeren and had hurried off on the sixty-mile journey. He found that one goat had already been eaten but he was in time to save the second from the pot and it now prowled happily round his yard devouring the weeds. A few days later he transported the serow twenty miles north to Ketambe and ferried it across the fast-flowing Alas River. The animal was in no hurry to leave its saviour but finally roamed off into the forest when the boat bobbed back across the rapids.

It was here at Ketambe that Herman was planning a rehabilitation centre for the release of captive and illegally owned Sumatran orangs. Under the direction of a local *pawang* a bungalow and animal shelters were already half built. In many ways the set-up was similar to that at the Sepilok in Borneo. Ketambe, however, had two great advantages. Since the site was across the river from the little-used road to Blangkedjeren it was remote from any human interference and, conversely, the orang-utans could not cause havoc in the *kampong* gardens. Moreover, the area abutted on to a vast tract of over a million acres of forest, inhabited by a healthy population of wild orangs. Herman already knew several of the residents and scattered 'villages' of nests attested to the presence of more. Game rangers had hacked a convenient system of tracks through the region and constructed a treetop hide overlooking a waterhole. The wet sand bore the imprints of deer and a small cat and only a week or so before Herman had found tracks of rhinoceros a little way downstream.

The forest here was much taller and wetter than round my camp on the Ranun River. Everywhere abounded the terrible *Latang*, whose stinging leaves cause weeping, red weals wherever they touch the skin. These wounds smart for days and are especially painful if they come into contact with water. Since there were no elephants to clear trails, travel through the thick undergrowth was difficult and there was certainly no shortage of leeches. I did not envy Herman his choice to work in such a place but the area was rich in fruits and vines and ideal for orang-utans.

Farther south at the village of Balelutu we visited three orangs waiting to be moved to their new home. The youngsters had destroyed the henhouse in which they slept and were gradually expanding their range to terrorise the neighbouring banana, pawpaw and coffee plantations. In spite of their years in captivity they were capable nest-builders and sampled the wild trees as readily as the cultivated fruits. I could see no reason why they should not readily adapt to normal living in the reserve. The eldest of the trio was an adolescent female and the undisputed ruler. Her companions were only infants and clung together for mutual comfort at every opportunity.

The road back bumped and twisted along the valley. It was in an appalling condition and used by little traffic. After every flood all the bridges had to be re-erected. Even on the main Sumatran highways road repairs were a pretty slapdash affair. Gangs of coolies poured basketloads of rocks into the gaping potholes and the surface was levelled off with thick, red earth. At each approaching vehicle work stopped so that the bouncing transport could compact the stretches already completed. It was a great pity that after so much back-breaking toil the very next heavy rainstorm undid all their efforts. Only the most important trunk roads merited a cover of tarmac and so were spared some of the ravages of the elements.

We drove along the Alas valley enjoying the breathtaking view. Rising up to the misty mountains the forest was a beautiful blend of every shade and hue of green with here and there a brighter splash of orange or flame. Siamangs and leaf-monkeys, completely untroubled by our presence, performed by the wayside and a band of macaques gambolled along the beach beside the river. Night overtook us as we meandered through the *kampong* lands, where water buffalo wallowed in the ditches and chickens and ducks paddled among the *padi*.

Back at Kutatjane Herman told us the history of the orphans we had met. All three had belonged to a policeman in Lau Baleng. Until recently one of the greatest difficulties of conservation in Sumatra has been the impossibility of enforcing the game laws. Some of the worst offenders are the police and army, who are all issued with firearms, and even if the Conservation Department did confiscate illegal pets they had nowhere to keep them and no funds for their upkeep. Now at last with money from international bodies, especially the Dutch, and the help of Herman and others the game rangers are in a strong position to be able to enforce the rules. Even so it would be extremely difficult to gain a conviction against a poacher and there is still a thriving trade in captured animals. It is considered fashionable in Indonesia to own a tame orang-utan, siamang or monkey and many such unfortunate animals can be seen in gardens in the wealthier residential areas of Medan and Pematang Siantar. Europeans are just as much to blame as the Indonesians and even we, ourselves, had been guilty of buying Tigger. Hawkers on bicycles boast a selection of animals but only produce the goods when they are confident of a safe sale. Birds are offered openly in the bird market of Medan and multi-coloured finches flutter and trill in their crowded cages. Yet another outlet are the local zoos, which are very popular and require constant restocking due to unusually high mortality.

But the greatest threat to the wildlife of Sumatra, and indeed of all Indonesia, is the destruction of the forest for agriculture or timber. The highland rural communities practise a wasteful type of slash and burn cultivation. Extensive areas of jungle are cleared but since it is illegal to sell the timber the valuable wood is burned and wasted. In the clearings the villagers plant hill rice or maize and when these are harvested the land is used for tapioca and chilli peppers. After a few crops all the goodness is leached from the soil and it is left to revert to forest. Secondary shrubs and bushes grow quickly but it would take a hundred years to return to climax forest again and this is never allowed to happen. Every two or three years the scrub is burned to encourage the growth of new grass on which the cattle and

Kathy and Tigger

The author and Jamie with the red apes

Above A fine Sumatran male in Kuala Lumpur zoo
Below Mother and baby

Opposite A tiger takes a bath and, below, a mountain goat saved from the pot

water buffalo may graze. The end product of this tragic process of declining utility is a type of acid heath, expressively referred to as '*Blang*'. Millions of acres where lush rainforest once stood are now covered in sterile *blang*.

This is the only way of life that the rural highlanders know and their whole culture depends on it. For centuries they have been nibbling at the edge of the jungle, causing its boundaries to shrink farther and farther back into the mountains. The day will come when there is no more accessible woodland to slash and when that happens the shifting cultivators will have to adopt a new method of agriculture. Whether they are forbidden to destroy any more forest now or are allowed to continue their ancient practices until there is no jungle left, eventually they will be forced to change their methods. Nor is it mere sentimentality to want to preserve as much primary rainforest as possible. The oxygen in our atmosphere, on which we are so dependent, is released from photosynthesising green leaves, which at the same time bind up harmful carbon dioxide into useful sugars as the first step in long food chains, many of which man exploits. The great forests of the world do at least half the work and we should be in grave trouble without them.

Problems arising from large-scale destruction of forest to extract valuable timber are rather more complicated. No-one can expect a poor country to ignore such an important resource for the sake of a few rare animals. While the total rape of vast areas for a quick profit is absolute madness in long-term ecological and economic terms, many governments do make attempts to follow this up with land redevelopment schemes to improve the lot of the local people. In North Sumatra the authorities have tried to find a workable compromise by allowing selective timber felling under licence. Properly controlled and where the cutting rate equals, or is less than, the rate of tree growth, such a management policy can be both profitable and successful in conserving the variety of habitats necessary to support a rich fauna.

To see just how effective this solution was, Herman, Kathy and I planned to visit a low-lying reserve to the north of Medan where a Chinese timber company had been operating for several years. We wanted to see what influence the human activity was having on the animals of the region. In Sikundoer we should be able to compare three forest types: virgin jungle visited only by *jelutong* tappers, areas where timber pirates had been felling illegally for many years but on a small scale and using only manual labour and, lastly, forest where a big company was conducting more thorough logging and bringing in heavy machinery for towing.

The Land-Rover was loaded up for the trip and we headed for Pematang Siantar, where Ans would stay with friends and look after the baby orang-utan, who was still weak and ill. Our way took us past beautiful Lake Toba and we picnicked among the pitcher plants on a high crag overlooking the shimmering, blue waters. Hemmed in on three sides by glowering black cliffs, the lake is fed by a magnificent waterfall, which plunges down in a single, roaring sheet of white foam. Far below, the *padi* fields, yellow and green in the afternoon sunshine, snuggled at the foot of the mountains and on the distant southern shore we could see the hazy outlines of the white villas of Prapat, a flourishing tourist centre. Resisting these temptations, we wound on to Pematang Siantar, pleasant, cool and with an unmistakable Dutch flavour about its architecture that recalled colonial days. The Sunday *pasar* (market) was in full swing and we wandered among the crowded stalls, laden with everything from exotic fruits to wickerwork baskets and gay *sarongs*. Here we purchased a huge pile of chilli peppers and salted fish, provisions for the expedition.

Bidding farewell to Ans we drove north through groves of *rambutans* and tea plantations to Medan to collect two of the conservation officers who were to accompany us and then on up the coast to Sikundoer. We were to be the guests of the two Chinese brothers who held the timber concession and were to travel up-river on a returning company barge.

We arrived at the river to find the *kampong* idling in the midday sun. The river was tidal and the water not yet deep enough for the heavy barge to navigate the shallow sandbars. During our wait we amused ourselves by following a troop of *lutong*, silver leaf-monkeys. One of the females carried a tiny baby, bright orange in colour in startling contrast to the charcoal and silver coats of its parents. A deep throbbing rent the peaceful air and the cumbersome craft chugged into view. At once the sleepy hamlet sprang to life. Ropes were flung, instructions yelled and the great boat floundered in to the bank and was made fast. Half-naked youths appeared from nowhere and swarmed over the barge, manhandling the rough-hewn planks on to a veteran furniture lorry. Our Chinese friend was everywhere, hanging on to the truck's footplate, gesticulating wildly as the vehicle churned backwards, offering cigarettes to the relaxing helmsman and ensuring that a continuous flow of cups of tea moved in our direction. The women came down to the muddy water to watch the fun and do their laundry. Dripping *sarongs* were generously larded with soap, pounded on the wooden landing stage and rinsed in the murky brew. Amazingly, they emerged clean and bright.

After an hour's frenzied activity the cargo was unloaded and stores taken aboard for the logging camp up-river. Huge, open baskets were packed to the brim with onions, rice and enormous packs of cigarettes, for, at sixpence for twenty, the merchants could afford to be generous. We picked our way carefully across a narrow plank over the gaping hold and settled as comfortably as possible on the oily bulwarks. The engineer tugged and heaved and juggled with a spanner until the ancient engine throbbed into life, belching out thick clouds of smoky fumes. Water swirled, stragglers leapt shrieking overboard and we steamed ponderously out into the middle of the stream. Small shacks ranged along the river banks and naked, brown children splashed at the water's edge, squealing with delight at our progress. Soon we left the *kampongs* far behind and dark jungle overhung the lazy current. Fig trees trailed sweeping branches in pools where shoals of silvery fish darted and fled from our wake and, high aloft, bees swarmed noisily round their pendulous nests.

A small *prahu* rounded the bend in front of us, bearing three naked urchins. At sight of Kathy they pulled hastily into the far bank and modestly donned their patched shorts. Waving and giggling, they paddled past but after a few hundred yards shipped their oars and removed their clothes again, before proceeding on their way. Soon we were chugging into a small landing where the forest had been cleared. A large timber lorry was waiting. Our camping gear was thrown quickly on to the lorry and as we clambered after it the driver started off.

The truck clattered along the rutted track between banks of bracken and yellow *simpoh* flowers. Beautiful butterflies with wings of soft browns and iridescent purple accompanied our passage. On we bumped over fallen trunks, up and down steep hills, the vehicle at times threatening to bellyflop into the valley beyond. We were flung from side to side and clung to our precious cameras for dear life, terrified lest they should fall from the open sides and be crushed by a giant wheel. When the terrain became too difficult for the sliding lorry the running boys leaped from their perches behind the cab. Their grease-covered shirts wet from the effort, they fixed up winches and heaved and hauled, yelling encouragement to the driver. At last, wheels spinning, engine screaming, the juggernaut strained against the taut wires and slowly mounted the slippery hillside. As we roared off down the next descent the runners swung aboard again without any care for life or limb. In this fashion we approached the main timber camp, an extensive settlement of ramshackle wooden buildings and the great cutting sheds, where saws whined as the huge logs moved through.

We spent a week in Sikundoer, staying in a small *pondok* a few miles from the main logging centre. Our host provided us with two boys to look after our needs and a *pawang* as a guide. Unfortunately, the latter had definite preconceived notions of what was expected of his role. To ensure that we were getting our money's worth he would lead us, for the first mile or so of each day's trip through unpleasant swamp, thick tangles of jungle and over precarious thin saplings bridging deep chasms. Kathy lost her nerve on one particularly pliant trunk and, as it swayed and sagged beneath her, sat down hurriedly and inched the rest of the way on her bottom, much to the amusement of our sure-footed guide.

One morning a display of tree-felling had been arranged. A lithe youth cut two saplings and bound them firmly together to form a V-shaped platform round the trunk of the chosen giant. Balancing with bare feet on these narrow supports, he wielded his axe vigorously. The sharp blade bit into the rough bark but each blow made little impression. This was going to be a long job and we were amazed that they employed such primitive methods. The young woodcutter obviously knew what he was up to for when he had removed a thick slice from one side of the trunk he moved position and started afresh. He was drenched with sweat from his exertions and paused every few minutes to wring out a dirty rag tied round his waist and rub himself down. It took almost an hour of strenuous effort before the huge tree trembled and began to fall. With a final blow the lumberjack jumped clear and the mighty giant crashed in slow motion to the floor. We paced it out and found the trunk was nearly two hundred feet long, three times the height of a mature English oak.

Several neighbours were dragged down by the weight of the falling boughs and from one of these a russet flying squirrel launched itself to find a more reliable home. It glided for about sixty yards, alighted, scrambled clumsily up the bark and swooped off again. Arms outstretched and long, busy tail streaming behind, it appeared like a big brown kite floating serenely from tree to tree. Kathy was thrilled by this unexpected daytime performance made, she was sure, especially for her benefit.

While in Sikundoer we investigated areas that had been subjected to varying amounts of human disturbance and ended our survey by visiting the camp of timber pirates, who were operating without licence only a couple of miles upstream of our own *pondok*.

A high bank sloped away steeply, giving the impression of a ski-run without snow. At the top three slim, brown youngsters puffed and grunted as they hauled a huge log to the brink. These were the timber pirates. They could not have been more than twelve or thirteen years old and we never saw one of their number aged more than twenty. With their long dark locks and coloured headbands they looked like Aztec Indians. A thick rope was tied round the great trunk and they tugged and heaved at the rusty winch, working with intense fury to move the timber a few painful feet. It took them more than a month to drag one giant tree from where it fell to the water's edge. The precious booty rolled the last hundred yards downhill to splash into the stream and wait with its fellows for a flood to wash them down to the main river, where they would be collected by their illicit plunderers. Since their route lay past the Chinese sheds we suspected that many would never reach their destination but would end up on the timber barge. Several logs already lay in the muddy water, not in the neat rafts that one associates with Canadian lumberjacking, but in a confused heap, overlaying one another and entangled with the vegetation. Obviously the timber was very valuable for the pirates to toil under such difficult conditions when they risked losing much of their loot to the Chinese company or projecting sand spurs. Yet relations between the two factions seemed good and I imagined they must have reached some mutually beneficial agreement.

In spite of the extensive timber operations in Sikundoer we had seen or found evidence of a wide variety of animals: orang-utans, siamangs, gibbons, monkeys, squirrels, hornbills, flying lemur, deer, pigs and elephants. But although these species were present in the area they seemed less numerous than in other parts of Sumatra I had visited. Even the undisturbed sectors of Sikundoer seemed poor in wildlife. A sure indication of the diminished fauna was the scarcity of leeches on the acid, swampy soil but they were old acquaintances we were happy to miss.

Where the pirates had been working, slowly but consistently for long periods, the forest had been opened up and the wide, log-rolling trails made complete breaks in the available arboreal routes. Gibbons had survived, leaping the gaps, and flying lemurs and squirrels were now at a distinct advantage but the siamangs had retreated to virgin jungle. Orang-utans do not mind descending to the ground to traverse clearings and we found several old nests in logged areas but no sign of their builders. Groups of leaf-monkeys and crab-eating macaques ranged the river banks and the open glades supported a wonderful selection of birds. I was intrigued by the racket-tailed drongos and spent a mosquito-ridden evening squatting in a hide to obtain one distant shot of this elegant and harmonious songster and his curious adornments.

Animals seemed to have deserted the areas where the timber company were currently felling but there was plenty of evidence to suggest that they returned to recolonise recently worked parts once the noisy machinery had departed. Since our Chinese friends had a lease for a limited period they moved through the forest extracting only the most valuable wood so that much of the canopy was left standing and the region should quickly recover.

Fortunately the trees which give the best timber are not necessarily those that produce the main foods of the forest wildlife but the toppling giants drag down many important vines, epiphytes and strangling figs. Moreover, the trails and clearings left in the wake of felling operations make the region more accessible and the fauna more visible, important considerations when assessing the value of a nature reserve. But although I could see that many species did not suffer unduly, or even benefit from this human interference, the orang-utan did not seem to have fared so well. Orangs are extremely sensitive to disturbance and while some had returned briefly to nest in logged areas none seemed to have made this their permanent home. Instead the red apes had fallen back farther into the forest and I could judge what effect this would have on the reproductive activity of the population from my Bornean findings. Orangs are slow breeders and it would take many years for them to recover and recolonise their traditional haunts. Indeed I doubted if they would be given sufficient time to do so before timber-felling operations cut farther into their range. The pace of commercial expansion is simply too fast and the orang-utan seems the inevitable loser.

My own time, too, was running short but I had one last important matter to attend to. Dragged from one leech-infested jungle to another, Kathy had hardly had the honeymoon of her dreams and goodness knows what she might tell her mother on our return. At Medan we dispensed with our *parangs* and heavy kit, then travelled east through Java to the paradise island of Bali. A fortnight of fresh coconuts, palm-fringed beaches, Hindu temples and colourful festivals should soon sweeten memories of steamy forest and the strange, red apes that so fascinated her crazy spouse.

Epilogue

It is only in the field that one can collect the information necessary to achieve a good understanding of wild animals, their real problems and the solutions they employ to overcome these. Equally, however, it is only when one takes time to organise, analyse and sort through these data that a clear picture emerges about the animals' habits. This was the task I had left for a year's residence in Oxford, writing up my doctorate thesis.

The most outstanding feature of orang-utan life was its solitary nature. This had impressed me everywhere. Females were accompanied only by young offspring and males usually ranged alone. Adult animals did not link up with others of their sex and males consorted with females solely for mating purposes. Whereas his Bornean counterpart indulged in a brief partnership and then left the female to rear the young alone, the Sumatran male tended to stay with his mate until after the birth of the new baby and played a greater part in family life. His fierce presence was probably necessary to protect the vulnerable infant from aggressive and competitive siamangs and savage predators like tiger and panther, enemies not found in Sabah. Apart from this greater paternal involvement, however, Sumatran orangs resembled their Bornean cousins in leading a life of solitary independence.

Among monkeys and apes such a lack of sociability is unique but, in view of the orang's great size, relative immobility and mainly frugivorous diet, it is ecologically very sensible. Since orangs are slow-moving and cover little ground in a day it is difficult to imagine any area rich enough in fruit to support a big party.

Some of the animals seemed to be shifting, forest nomads but many lived in large, but definite, home ranges. They might leave these in months when food was scarce and neighbouring regions more productive but they soon returned to their old haunts. These ranges were not exclusive, like the territories of gibbons and siamangs, and many orang-utans resided in the same area and often met at good feeding sites. Such meetings were usually peaceful, with only the big males showing jealousy, defending their domains to keep out strangers and other high-ranking rivals.

After sixteen months in Borneo I was still not certain what sort of relationships existed between animals sharing a range. They seemed to tolerate one another, yet never showed any signs of friendship or interest in their neighbours' affairs. I suspected that these were all members of dispersed, loosely banded groups with the calling males acting as group leaders. My Sumatran findings supported this theory for, although the animals were foraging alone, they were more closely clumped with concentrations centred around single, vocalising males.

It was these old males who posed the greatest problem. What is their role in orang-utan society and how do they benefit from their strange behaviour? Undoubtedly it is to the advantage of all the residents to have the group range guarded against intruders and other aggressive males. Moreover, they must benefit from following the challenging leader with his greater experience and knowledge of likely food sources. But what of the males themselves? Altruism is a concept few zoologists will accept and we must discover what advantages the loud calling bestows on the individual performer.

The need for male-male spacing, the threatening nature of their displays and calling behaviour and the whole sexual adornment of the adult males, their great bulk and size-exaggerating features, such as face flaps and long hair, all suggest that intense rivalry and competition exists between males. But competition for what? The most obvious answer and the opinion held by the Bornean Dyaks is that they are competing for females. If males attracted receptive females by their calling this would account for the sexual dimorphism and the inter-male rivalry. Females did occasionally move towards callers but far more often they avoided such potentially aggressive animals, hiding quietly till they had passed or moving away themselves. In fact, it was the less well-endowed, less frightening sub-adults who seemed to have the most luck with the ladies.

In zoological terms the success of an individual is measured by his ability to leave descendants. This does not mean merely how many offspring he can father but also how likely these are to reproduce in their turn. They, too, must have sufficient status, food and particularly space to breed.

It is here that the answer to the male's vocalising seems to lie. By advertising his presence and general bad temper he may spoil his own chances for further reproduction but he also disrupts the mating of his neighbours. He is in effect producing a vacuum, maintaining a range where the orang population is below the carrying capacity of the environment. In this way he guarantees that his offspring have sufficient space, and therefore food, to grow up and breed after him. This is his best tactic for promoting his own lineage.

Thus the male's breeding strategy seems to be divided into two distinct phases; as a sub-adult he fathers as many offspring as possible and as an adult he consolidates and protects his contribution to future generations. As he grows older he becomes larger, uglier and more frightening. This makes consorting more difficult but enhances his effectiveness as range guardian.

Probably the old males once played a much fuller part in orang-utan society and it is only recently that the sexual advantage has passed on to the sub-adults. Geological quirks have preserved large numbers of fossil orang-utan teeth, some dating back to the Pleistocene era, over half a million years ago. These teeth are much bigger than those of present-day orangs, suggesting that the prehistoric animals must have been twice as bulky as their descendants. Since big orang-utans to-day find travel through trees difficult their giant ancestors must have been even more terrestrial. They probably ranged on the ground, like modern gorillas, in large bands, protected by enormous males. Unlike their present equivalents, such patriarchs could have lorded it over their subjects and had the pick of the females in their group.

These ancestral orangs occupied the woodlands and montane forests of southern China but a cold spell during the Ice Age forced them south into the tropical regions of what is now Indonesia. At that time there were two land bridges, linking Java with Malaya, via Sumatra, and with China via Borneo, the Philippines and Formosa. When the climate warmed again the lands to the north became suitable for orangs once more, but the rising seas, swollen by melting ice, had submerged the connecting bridges, leaving the red apes stranded on the islands of Borneo, Java and Sumatra.

The apes that had evolved in semitropical woodlands now found themselves restricted to tropical rainforest, where food was confined to the continuous canopy that also provided arboreal pathways free from predators. With selection favouring animals of a more comfortable size for treetop life, subsequent generations of orang-utans became gradually smaller. These more agile forest orangs have found community living no longer necessary, and even disadvantageous, so that groups have become dispersed in the curious way we now find them.

Had a change in habitat been all that the orang had to face all would have been well, but during the Pleistocene another hazard appeared in Indonesia, Stone Age Man. In Java Early Man spread quickly. The fertile lowland plains, and eventually the uplands, became populated and cleared so that the apes were forced back on to the isolated volcanoes that still sported forest and soon became extinct.

The more mountainous islands of Borneo and Sumatra were far less attractive to man and his inroads here have been much slower. Charred remains from the Niah caves in Sarawak show that orangs featured commonly on the menu of Stone Age Man and the apes must have been an easy prey to spears and blowpipes. As competitors for the same types of fruit and as hunter and hunted, man and orang-utan have had a long and violent relationship. Wherever man has penetrated the orang has vanished, hunted for the pot, caught for pets or simply because his habitat has been destroyed. Whether this process will continue to the eventual elimination of the red ape, as in Java, is still in our hands. There are sufficient animals and enough forest left for this precious species to survive but time and human greed run hard against the orang-utan. He has survived the predation of the forest tribesmen, the nineteenth-century naturalist collectors and sportsmen and the awful trade of poached babies but he stands no chance when his native forest is destroyed either for cultivation or for timber.

Until recently timber extraction in Sumatra has been operating on a small scale. Men with handaxes fell the trees and water buffalo drag out the logs. Power-saws and lorries are replacing these old methods but the greatest threat to the orang in Sumatra is still the expanding human population with its wasteful slash and burn practices of cultivation. The steepness and inhospitality of the terrain in the Gunung Leuser and Langkat Reserves are probably the animals' best protection. In Borneo, however, completely the reverse is true. The human population is relatively small but timber is one of the country's main resources and the navigability of the great rivers facilitates its removal. Logging in Kalimantan and Sabah is continuing at a frightening rate, too fast for the orang-utan to cope. Unfortunately, the Ulu Segama, where I had found so many red apes, is scheduled for timber concessions and, despite the efforts of de Silva and others, Sabah has no large sanctuary. Sepilok and the Mount Kinabalu National Park probably support too few wild orangs for them to be viable in the long term. Reserves in south and east Kalimantan at present provide sanctuary for healthy populations of orang-utans but even these are being encroached upon by logging companies.

A year away from the forest is a long time but we had not been idle. I had produced a thesis and Kathy a son. At the age of three months baby Jamie was whisked off to the jungles of Malaya and from there it was only a short hop to Sumatra. During our brief visit we were able to renew old friendships and see how things had changed in our absence.

When I made a sentimental pilgrimage to my old camp at Ranun, I was apprehensive to discover that another vast block of forest had been cleared for cultivation and shocked at the speed with which the *kampongs* of Lau Djohar and Lau Ranun had expanded. Now there was so much traffic that the villagers had even built a bridge across the river. Where my old hut had stood there lay only a pile of rotting timber but here, at least, the forest and its monkeys were just as I remembered. I surprised a family of elephants at the salt caves and at last managed to photograph one of the shy giants. On my return I met my friend Hitam, the enormous lone siamang, still leading the life of a social outcast. I found no orang-utans in the vicinity but a fig tree was thickly covered in half-ripe fruit and I was sure that Co and his followers would arrive in the next few days. I was greatly tempted to wait and see them but had to hurry on to Atjeh to visit the Rijksens in their new bungalow adjoining the orang-utan orphanage at Ketambe.

Jamie was duly introduced to the sixteen young orangs, who now wandered freely in the jungle surrounding the Rijksens' camp. The orang-utans were fascinated by this tiny, white ape and came to inspect, smell and kiss the delighted child. Jamie gurgled appreciatively at this friendly attention, unaware that he was in imminent danger of being carried off to a leafy nest.

It was wonderful to watch these beautiful animals moving at ease through the treetops, feeding on the leaves of the mimosa vines and making their nests for the night. Many of the animals had been rescued from unhappy captivities; one near-adult female had spent a year locked in the cramped space of a disused car and another orang had been badly burned by an electric cable. Now they could enjoy the best of both worlds, the freedom of the forest and the knowledge that they would never starve; rice and pawpaws were always available at regular mealtimes should they be needed.

We were delighted that the scheme was running so well but it was not only with the rehabilitated animals that Herman was having success. He had seen many wild orangs in the area, too. Some had shown much interest in the newcomers and lingered several days near camp to watch the proceedings. I was sad to have missed them and very jealous when Kathy saw the only wild orang-utan of our stay while I was away.

Elsewhere prospects for the orang-utan did not look so happy. We were dismayed by the speed at which human settlements were spreading throughout North Sumatra. All along the Alas valley bare *ladangs*, newly planted, stretched beside the river, where just fifteen months earlier had stood green jungle. Even within the boundaries of the reserves new timber concessions have been granted and huge machinery was destroying the orangs' home at an unprecedented pace. How long can it go on? Will anyone have the sense or the authority to call a halt? I hesitate to answer. Having finally emerged from two decades of political and economic troubles, one hundred and thirty million Indonesians are a large force to control and their desire for rapid progress is completely understandable. The orang-utans will certainly lose a lot more ground. We can only hope that those impressive limestone mountains that have served to protect them until now can withstand this latest wave of human expansion.

The orang-utan problem is not insoluble. The red ape can be saved but it will require a considerable amount of money and international goodwill to secure the necessary steps and establish real safeguards for extensive tracts of forest, capable of supporting large, wide-ranging and reproducing populations. It is difficult to measure the value of this unique animal against the more obvious advantages of commercial gain but it is surely time we paused to consider before we eliminate another species, a fellow ape, our cousin Mawas.

Index

Index

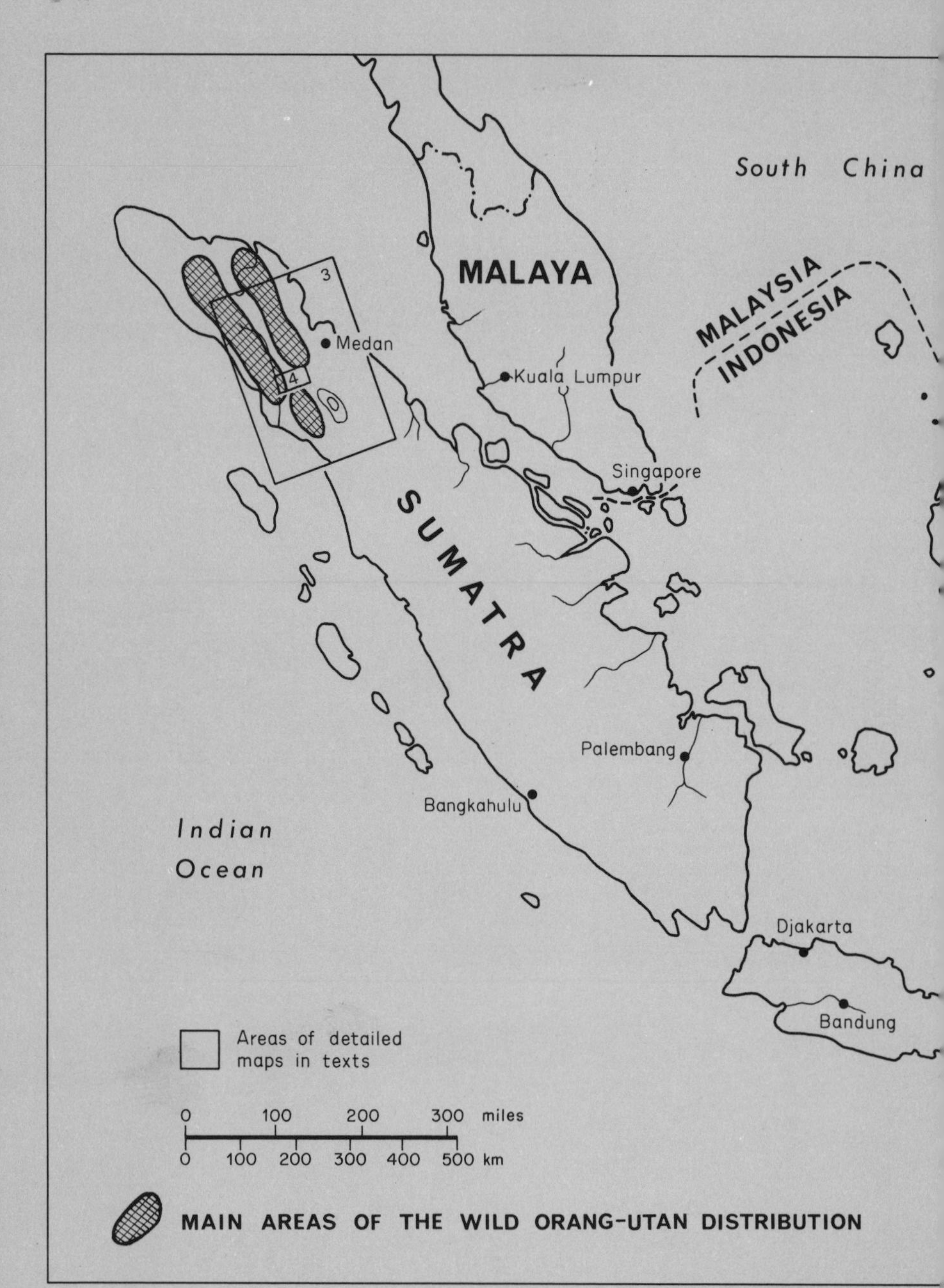

South China
MALAYA
MALAYSIA
INDONESIA
3
Medan
4
Kuala Lumpur
Singapore
SUMATRA
Palembang
Bangkahulu
Indian
Ocean
Djakarta
Bandung
Areas of detailed maps in texts
0 100 200 300 miles
0 100 200 300 400 500 km
MAIN AREAS OF THE WILD ORANG-UTAN DISTRIBUTION